THE CITIZEN'S GUIDE TO LEAD

UNCOVERING A HIDDEN HEALTH HAZARD

BARBARA WALLACE AND KATHY COOPER

for the
NIAGARA NEIGHBOURHOOD ASSOCIATION

NC Press Limited
Toronto 1986

Photo Credits
Simon Glass: front cover; pages 3, 16, 23,27,31,32,38,42,43,53
Laura Jones: page 78

Canadian Cataloguing in Publication Data

Wallace, Barbara
The citizen's guide to lead

Bibliography: p.
Includes index.
ISBN 0-920053-92-0

1. Lead - Environmental aspects - Handbooks, manuals, etc. 2. Environmental pollution - Handbooks, manuals, etc. 3. Lead - Physiological effect - Handbooks, manuals, etc. I. Cooper, Kathy. II. Title.

RA1231.L4W35 1986 363.7'384 C86-094336-4

We would like to thank the Ontario Arts Council and the Canada Council for their assistance in the production of this book.

New Canada Publications, a division of NC Press Limited, Box 4010, Station A, Toronto, Ontario, Canada, M5H 1H8

Printed and bound in Canada

Distributed in the United States of America by Independent Publishers Group.
One Pleasant Avenue, Port Washington, New York, 11050.

To all women and men whose diligent efforts over the years are finally focussing public attention on the dangers of low-level lead exposure.

Contents

Measurement Units

Frequently-used Measures

kg	kilogram	$\mu g/m^3$	micrograms per cubic metre
g	gram	mg/m^2	milligrams per square metre
mg	milligram	$\mu g/dl$	micrograms per decilitre
μg	microgram	ppm	parts per million
L	litre	ppb	parts per billion
dl	decilitre		

Equivalent Measures

1000 g	= 1 kg	1000 ppb	= 1 ppm
1000 mg	= 1 g	1 $\mu g/g$	= 1 mg/kg = 1 ppm
1000 μg	= 1 mg	1 $\mu g/kg$	= 1 ppb
1000 kg	= 1 tonne	1%	= 10,000 ppm

See Appendix F for more details on measurement units.

The Importance of Very Small Amounts

A concentration of one part per million may seem very small. It is the equivalent of only one gram of salt in a thousand kilograms of sugar. But, such very small amounts can have very important effects depending on what the substance is. For example, less than two parts per billion of diethyl stilbestrol (or DES) is capable of causing cancer in mice; or the lack of only one part per billion daily of Vitamin B-12 can result in serious health problems in humans. It takes only about one-quarter of a milligram of lead every day to cause very serious lead poisoning in a human adult. To avoid health problems, a child's daily intake of lead should be no more than 100 micrograms (1/10 of a milligram or 1/40 of a level teaspoon).

Acknowledgements

We gratefully acknowledge the significant contributions over the last twenty years made by a large number of unnamed scientists, governmental officials, commentators, and private citizens whose efforts have clarified the lead issues that are affecting us all.

To all the professionals and lay people who volunteered to read drafts of the chapters and whose helpful comments have provided us with valuable insights that have added much to the quality of the final product, we are especially thankful. The primary readers and reviewers are: Norm Bates, Marjorie Cooper, Chester Duncan, Robert Elias, Ed Ellis, Pamela Hickman, John Jackson, Sue Kaiser, Joanna Kidd, Maureen McDonnell, Barbara McElgunn, Anne Moon, Jerome Nriagu, Pat Perkins, James Pirkle, Elizabeth Quance, Michael Rachlis, Ted Schrecker, Lesbia Smith, Toby Vigod, Ray Vles, Milton Wallace, and Jocelyne Wallingford.

We owe a profound debt of gratitude to all those who have patiently given us valuable assistance in understanding the professional complexities of environmental, medical, technological, economic, and legal matters related to lead: Paul Aumuller, Chester Duncan, Robert Elias, Ed Ellis, Darryl Hogg, Marcus Hotz, Tom Hutchinson, Peter Irwin, Ira Kaufman, Scott Marsh, David McKeown, George Nagy, Jerome Nriagu, Bob O'Brien, Don Ogner, James Pirkle, Michael Rachlis, Rob Rinne, Joel Schwartz, Barney Singh, Lesbia Smith, and Ted Yates. Any errors in the interpretation of the advice we received from these professionals or any other errors in this book are solely the responsibilities of the authors.

We acknowledge with thanks permission from the Federation of Ontario Naturalists to reprint the *Environmental Bill of Rights* from their *Hazardous Waste Educational Resource Kit;* from the CBC-Radio "As It Happens" show to reprint *Dying of Lead;* from Ed Ellis of the Toronto Department of Public Health to use material from "Dustbusters" and "Safe Lead Removal"; from the staff of the Niagara Street School for use of a classroom for photographic purposes; from many parents for the opportunity to photograph their children; from those individuals whose lives have personally been disrupted by lead who graciously allowed us to tell their stories anonymously; and from the Sanderson Library in Toronto for storage of the Community Lead Impact Study's resource collection (including the non-published items cited in various references).

We are very grateful to the Niagara Neighbourhood Health Centre for their generous contribution of office facilities; and to the Lead Committee of the Niagara Neighbourhood Association for their constant support and encouragement that has in many ways made this book possible.

Special thanks to Alan Coulson for his illustrations and to Simon Glass and Laura Jones for their photographs. Their visual aids play an important part in communicating and personalizing the complex subject of lead.

This book has become a reality in large part thanks to the financial assistance of the City of Toronto and the Ontario Arts Council.

Thanks also to Caroline Walker, Ruth Chernia and Lisa Dimson of NC Press for their careful layout and editing help.

We are deeply grateful to our husbands Milt and Stu, and Barbara's children Elizabeth and Tal, who have given us the freedom to complete this work and sustained, encouraged, and helped us throughout.

Introduction

This book was written because people need to know more about lead. Although few of us realize it and relatively little public attention is being focussed on it, current levels of exposure to lead are causing serious health problems now. This lack of attention to lead (which may be due to the fact that it is such a familiar substance) is in sharp contrast to the widespread public concern over exotic pollutants, such as PCBs and dioxin. While these exotic pollutants can cause very serious health problems, for most people they are little more than risks that have a low probability of occurrence.

Because of this lack of public concern and strong lobbies by the lead and petroleum industries, governmental lead controls (particularly the amount allowed in gasoline) are weaker in Canada than in the United States and most other industrialized countries. The degree of exposure to lead in 1986 in Canada is high enough to cause health problems for thousands of Canadian children and to subject possibly half of our children to a slight degree of abnormal functioning in some of the cells and systems in their bodies.

A decade or two ago, lead was not considered harmful unless the amount was large enough to cause "lead poisoning," a condition that can involve severe brain damage or death. Public concern over lead issues peaked in the mid-1970s when the media began focussing on cases of lead poisoning in children living near urban lead industries. Some of this story and the flavour of that time is clearly evident in the transcript of the CBC programme "Dying of Lead" that can be found in Appendix A.

At about the same time and continuing until the present, extensive scientific research sought and found clear-cut evidence that low levels of lead in the body can cause significant damages to human health. One of the most important of these health effects is the lowering of intelligence in young children at relatively low levels of exposure to lead.

The amount of lead in the blood that is considered "elevated," that is, too high to avoid health problems, has been progressively lowered over the past 15 years from 80 micrograms per decilitre (μg/dl) to its current level of 25 μg/dl in children. Actual average levels of lead in blood in children not living in a lead "hotspot" area are about 10 to 12 μg/dl, levels clearly below the "elevated" level. However, averages do not tell the whole story and the "elevated" level does not take into account significant changes that can occur at much lower levels. For the average child there is an inadequate safety margin below the level at which health problems can occur, and for the many children with above-average levels, there is no safety margin at all. For the thousands of children with well-above average levels, health problems are occurring now.

The evidence is in. Low levels of lead in the general population are capable of causing significant health problems, and lead in gasoline has been identified as the major contributor to these lead levels. Economic analyses have shown that millions of dollars can be saved, primarily in automotive maintenance savings and children's health care costs, when large reductions/eliminations in leaded gasoline are made. It has also been demonstrated that leaded gasoline as an octane booster is not necessary in the vast majority of cars and, in fact, can cause significant mechanical problems.

Most industrialized nations have recognized these facts and taken steps to protect the health of their people by drastically reducing or eliminating lead from gasoline as quickly as possible. Given all these facts, it is not clear why the Canadian government does not remove lead from gasoline as quickly as possible, which is apparently the approximately two to three years that is required for the refinery industry to modify its operations. Instead, Canada is moving towards an unclear goal of "effectively eliminating" lead from gasoline over the next seven years. It is apparent that something other than avoiding health problems, saving money, and improving vehicle performance and operating costs is affecting the decision of the Canadian government.

The lead controversy that is occurring in Canada has very little to do with health effects or vehicle performance. It is a political controversy that is apparently motivated by the economic interests of those industries that derive profits from the use of lead in gasoline. These industries have applied considerable political pressure on the government to slow down the phase-down and phase-out of lead in gasoline.

Political pressure from one side can only be balanced by political pressure from the other side. The other side in this case is all the rest of us, people like you and your family and friends. You need to know the story that is in this book. We all need to know what is happening to us, our children, our vehicles, and our tax dollars. We need to know about other sources of lead exposure at work or in our neighbourhoods. We need to know what lead is doing to us, how we can protect ourselves, and what we can do to change the situation. Lead pollution is clearly an issue where the voice of the people needs to be heard.

The Citizen's Guide to Lead provides you with the information you need to protect yourself and your family from lead pollution as well as to act as a responsible citizen in the political debate. The *Guide* documents some of the health effects and problems. It describes how and where you are being exposed to lead and shows you what you can do to reduce these exposures to a safe level.

The book is divided into three parts. Part 1, *Highlights,* describes the primary health risks, the most important exposure sources, and personal actions all of us can take to reduce our risks. Part 2, *The Whole Picture,* besides going into more detail on most of these issues, includes important information on the history of human use of lead, the present degree of environmental contamination, and problems with and solutions for the most important governmental lead control strategies. In order to provide a complete story in Part 2, some of the material in Part 1 is repeated in a fuller context. Part 3, *Action,* provides details on what you can do to begin changing the situation. The appendices contain information that will be useful to many people with specific interests, especially those who might want to conduct their own "Home Lead Inventory." Since lead issues are intertwined in a complex fashion, an Index has been included to help you find all references to a particular topic. The case stories that are scattered throughout the book are based on actual people and incidents although the names have been changed to protect their identities.

The *Guide* was written in Canada for the Canadian situation. However, it includes considerable information on lead in the United States and can be used by people in all industrialized nations. In addition, people in developing nations will also find the information contained in this book useful particularly since industries that are losing leaded gasoline markets in the developed countries are exporting more leaded gasoline to developing countries, where there may be little or no regulation of lead.

One of the lead "hotspots" in Canada is Niagara Neighbourhood on the edge of downtown Toronto. This neighbourhood shares its living space with a secondary lead smelter, a major expressway, and a number of heavily travelled streets. For the past 15 years residents in Niagara Neighbourhood have been fighting for better control of lead. Parts of their story are told throughout this book. When the Lead Committee of the Niagara Neighbourhood Association became educated about lead health effects and exposure sources, they realized that they were not the only ones at risk. Out of their concern and interest for others, this book was conceived as the best way to put vitally important information into the hands of all the rest of us who are also at risk.

Barbara Wallace
Kathy Cooper
Toronto
July, 1986

PART 1 HIGHLIGHTS

Chapter 1

Lead Risks

"Carlos is going to be put in a special class, Mrs. Fernandes," said his second grade teacher. "He is not reading at all and he seems to be having some difficulty dealing with the classroom situation. He gets angry quite easily and shouts at me or kicks other students. Have you noticed any problems like this at home?"

"Oh, yes. I know what you mean. Carlos has always been hard to handle."

"Well, we've decided that he will do better in a different class. Our special ed class is set up with fewer students so that he can be given extra assistance. That should help him overcome his attention problems and let him settle down enough to learn to read."

Rosa Fernandes sighed as the teacher went on about how Carlos could be helped in the new class. She thought to herself, "Not again — is Carlos ever going to do anything right?" All week they had been having the usual fight at home to get Carlos to do his share of the chores or even to take a bath. And he was always so dirty. None of the other three children ever got as dirty as Carlos did. He was always digging in the ground and his mouth was so dirty, you'd think he was actually eating the dirt.

Last month there was that fire he had started in the basement. If the fire department hadn't got there in time, the house could have burned down. And now this — being put into a special class. Rosa knew it was for slow learners. And then next week she had to meet a social worker about the fire-setting incident. And the week after that, another blood test at the hospital to determine his lead level. None of her other children gave her all the problems Carlos did. Why was he so different?

Carlos had been put in the second-grade special class in 1974. Over the next few years, he moved from one special class to another, but never did learn how to read. Eventually he dropped out of school with no employable skills.

Carlos Fernandes did indeed have problems, and he would never be able to totally overcome them. His problems were the result of permanent damages he had received due to exposures to lead earlier in his life. There had been too much lead for his developing body, especially his brain. There had not been enough lead to make him feel sick, just enough to prevent him from developing normally.

The story of Carlos' exposure to lead began in the late 1960s and early 1970s, when he was two to five years old and his family lived near a lead smelter. During those years, all the young children in the neighbourhood had their blood tested for lead. Carlos' level was pretty high — about 40 μg/dl (40 micrograms of lead in each decilitre of blood). In the mid-1980s, this level would be considered high, but at that time only levels above 60 μg/dl were considered "elevated" or high enough to cause health problems. Over the intervening years, the "elevated" (or, looking at it from another angle, the "safe") level kept being placed lower and lower. But for Carlos, no matter what the numbers were, the amount of lead in his body prevented him from developing normally.

Carlos liked dirt and was always digging and playing in the soil in the backyard. Nobody at that time knew that the amount of lead in the soil was so high that playing in it could cause serious problems for Carlos. The lead in the dirt in Carlos' backyard, and in the dust in Carlos' house, got on his hands and under his fingernails and then into his mouth. From his mouth it went into his digestive system where about half of it was absorbed into his blood. And once it got into the blood, it could circulate everywhere in his body, including his brain and other organs and systems.

The lead that got into his brain caused a permanent change in the way he processed information. Brain wave patterns were changed and other electrophysiological processes (see Glossary) were affected. When he reached school age, these changes showed up as learning disabilities and behaviour problems. Some lead also got into other organs and systems in his body, which might explain why Carlos developed some long-lasting health problems.

Are children today being put into the same situation as Carlos was in the 1970s? Even if today's children are exposed to less environmental lead, are they safe or will they develop some of Carlos' problems? Are children living in cities, near lead industries, or near expressways at risk of developing some of these lead-based "developmental deficiencies?" What about adults? Is lead exposure a risk for them too?

The answer, unfortunately, for both children and adults is a qualified "yes." There are many children today that probably are showing or will show some developmental deficiencies due to lead in their systems. However, most of these children will not have problems as serious as Carlos did, since in most cases, children in the 1980s are exposed to less lead than those in the 1970s. Although adult problems are different from children's problems, adults too can experience health effects due to lead.

Figure 1: Young children are the population group most at risk for lead health problems.

Current levels of lead exposure for large numbers of Canadian children are too high to allow them to realize their full intellectual potential. Our present use of lead is forcing many of our children to experience more lead in their bodies than is good for them, thus diminishing the quality of their lives and health.

It is obvious that a lead-exposure situation such as this can cause small, but widespread, effects on the vitality and viability of our society. A growing number of historians and scientists believe that exposure of the upper class of the Roman Empire to fairly high levels of lead was a contributing, and perhaps large, factor in the decline of that civilization. Although the Romans knew about the poisonous effects of a single large dose of lead, they were not aware of other, less deadly effects from lower-level, long-term exposure.

Today we know about these other effects. While most people nowadays apparently are not exposed to anywhere near as much lead as the Romans, current exposure levels are high enough to affect the health of a fairly large portion of our society. These health effects could weaken many of us as individuals; and collectively, as a society, they could make it more difficult for us to deal with the complex and widespread problems we currently face.

In order to help you discover what should be done to reduce the risk of lead problems for you and your children, this and later chapters will provide answers to questions like these:

- what kinds of problems can lead cause?
- how did we get into this situation, anyway?
- how much lead is too much?
- is there too much lead in our children?
- where is all the lead coming from?
- how can I reduce my lead risks?
- how can we improve the control of lead?

Lead and the Fall of the Roman Empire

The ancient Romans used soft, non-porous lead for constructing water pipes, serving dishes, and cooking pots. They also boiled grape juice in a lead pot to produce a very sweet syrup, called sapa, which was used as a sweetener for wines and foods since cane sugar was unknown.

What the Romans did not know was that they were cooking up their own poison. Sweet and toxic lead acetate was formed when grape juice was boiled in a lead pot. Recent attempts to make sapa according to the ancient recipes have produced a syrup with 240 to 1000 parts per million lead. Since the aristocratic class was allowed more sapa than regular citizens or slaves, they were more seriously affected by lead.

It has been estimated that the aristocratic class absorbed about 250 micrograms (μg) of lead per day. Records of health and behaviour of the emperors in the two declining centuries of the Roman Empire are consistent with relatively serious lead poisoning. For example, Claudius, a glutton who loved wine, was dull-witted and absent-minded with disturbed speech, weak limbs, an ungainly gait, tremors, fits of excessive and inappropriate laughter, unseemly anger, and severe stomach aches. With a daily diet that included an estimated 250 μg of lead, it is no wonder the Roman emperors were unable to govern well. Many historians feel that these Roman lead exposures were a major cause in the decline of their civilization.

Estimates of daily absorption rates in 1986 for Canadian children living in cities are about 45 micrograms of lead and for those living near a lead industry in cities, about 70 micrograms of lead. Comparable adult values are 13 and 23 μg. Since these values are considerably smaller than the Roman level of 250 μg, our society does not appear to be in any danger of sliding into a lead-caused decline. However, the average daily absorption level for children is considerably higher than the adult level and is apparently exceeded by a large margin in a certain percentage of children.

Source: *Adapted from: J. O. Nriagu, "Saturnine Gout Among Roman Aristocrats: Did Lead Poisoning Contribute to the Fall of the Empire," New England* ***Journal of Medicine,*** *1983, 308 (11), pp. 660-663.*

What Kinds of Problems Can Lead Cause?

Lead is a poison and in large amounts (over about 80 to 100 μg/dl for children and over about 100 to 120 μg/dl for adults) it can cause death. It is very rare for anyone these days to have lead levels that high. However, even quite small amounts of lead can cause health problems. And almost everyone has more lead than people had in their bodies a century ago. Many people, especially children, have so much lead that their bodies are not functioning at their best.

Children have more problems with lead than adults because of their greater exposure, absorption, and sensitivity. For example, small amounts of lead in a fetus or young child can prevent the brain from developing normally. Lead in young children (especially those under six years old) can cause chemical changes in the blood, which affect many other organs, including the brain.

When measurements are made of the electrical activity in brain waves, changes can be seen at blood-lead levels as low as 15 μg/dl. Slightly higher blood-lead levels in children are associated with decreases in I.Q., learning disabilities, problems in focussing attention, slower reaction time, and behaviour problems.

Other types of problems in children are due to the fact that the body takes in lead and calcium similarly. If there is too much lead in the body, there may not be enough available calcium. Calcium is important for many bodily processes, including the development of teeth and bones and the transmission of information in the nervous system. Lead also interferes with the metabolism of Vitamin D, which results in abnormal functioning in many systems and tissues.

As mentioned above, adults have fewer problems with lead than do children. However, if an adult's blood-lead level is sufficiently high (see "How much lead is too much?" below), serious problems can develop. What may turn out to be the biggest problem for adults is the recent finding of a relationship between high blood pressure and relatively low levels of lead, especially in men. Because high blood pressure plays such an important role in two of the leading causes of death, stroke and heart attack, this finding is extremely important and is being investigated through further research.

Another major area of health concern for adults is the possibility of increased blood-lead levels after about age 50 to 60. All through adult life, most of the lead that is retained in adults goes into bones where it causes no problems. However, since the bone mass normally begins to decrease sometime between age 50 and 70, the lead that was stored there can come out of the bones and back into circulation in the blood. Relatively large amounts of lead in the bone could cause health problems later in life.

Other adult health problems at relatively low lead levels include changes in certain substances in the blood, slower conduction of nervous system messages in the arms and legs, reproductive problems, digestive problems, and some signs of kidney disease.

Symptoms of severe poisoning include abdominal pain, vomiting, clumsy movements, and weakness. This type of severe poisoning is now rare. The much milder forms of lead health effects seen today have correspondingly milder symptoms, such as headache, irritability, and tiredness, or no obvious symptoms at all. See Chapter 6, for a fuller discussion of health problems for children and adults and the blood-lead levels at which they can occur.

How Did We Get Into This Situation, Anyway?

Thousands of years ago in the times of cave dwellers, people had very low levels of lead in their bodies, because lead levels in the environment were very low. This very low level is called the natural background level, which, in human bones, is about one part per million (ppm). The first part (left bone and left tooth) of Figure 2 represents this natural background lead level. In those prehistoric times, most lead was still buried in the earth in ore deposits, often in combination with silver.

As time passed, metal-working techniques were invented resulting in the development of metal coins and decorative objects. As silver began to be mined in larger and larger amounts to produce these coins and objects, the ores from which it was extracted also produced quantities of lead. Experimentation with this now easily available lead revealed that it had several useful characteristics that made it a valuable metal in its own right: it was fairly soft so it could be easily worked; it resisted corrosion and breakage so that containers made from it were long lasting. Lead began to be used more and more, with the result that the amount of lead particles in the environment and in people's bodies increased. Recent measurements on the teeth of Peruvian Indians who lived about 1100 A.D. found lead levels of about 14 ppm.

This slowly-increasing use and environmental buildup of lead received a major boost at the time of the Industrial Revolution (beginning about 1750) and then another, even larger boost about 1925 when lead began to be used as a gasoline additive. As a result of these increases, present lead levels in teeth are about 188 ppm in people living in an industrialized part of the world. Even in the Canadian Arctic, Inuit people now have levels of about 56 ppm in their teeth.

Within the last century there have been many cases of lead poisoning leading to brain damage or death. Many of these cases resulted from occupational lead exposure. After the 1920s, lead poisoning began to occur less frequently in workers and more frequently in children. Children became exposed to lead through plumbing, older style kettles, and especially through chips of leaded paint.

Figure 2

Lead In People: Bones And Teeth

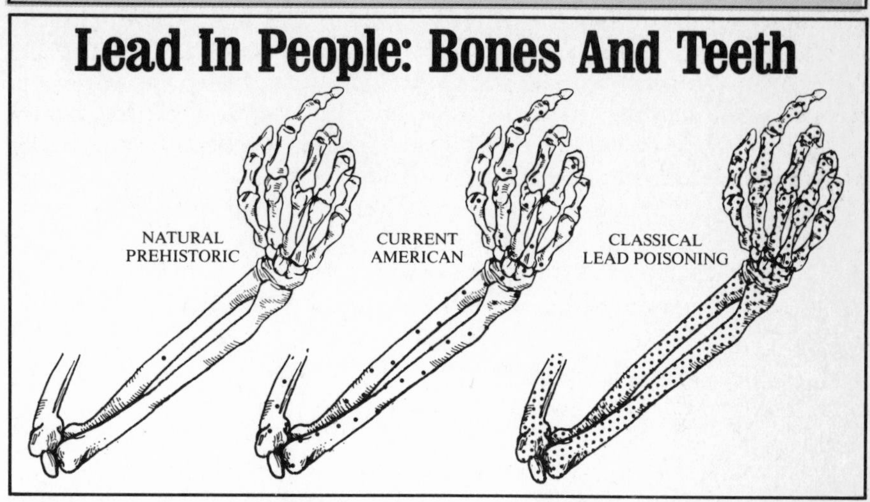

Comparative amounts of lead in people from different time periods, as measured in bones; natural prehistoric (one dot), average amount in present day Americans (500 dots), and the minimum amount which will cause classical lead poisoning in a significant fraction of a group of people (2000 dots). Each dot represents about 0.3 mg. of lead in a 70 kg. person.

Source: *Adapted from C.C. Patterson, "British Mega Exposure to Industrial Lead."* **Lead versus Health,** *M. Rutter and R.R. Jones (eds), John Wiley and Sons, Ltd., 1983, p. 18.*

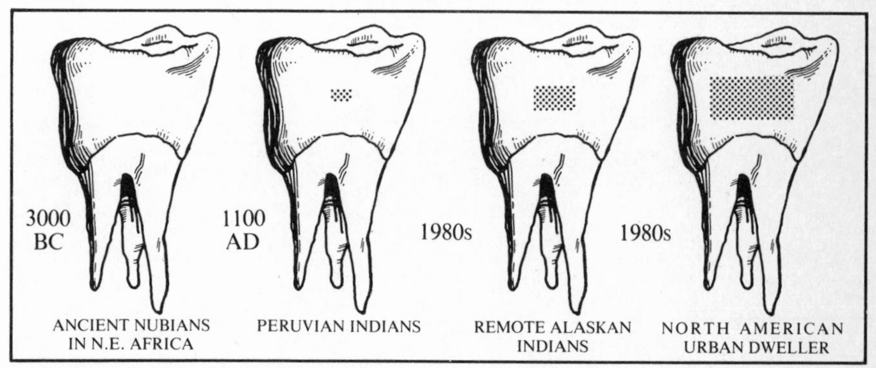

Comparison of lead in teeth from people who lived at various times in history including current measurements from remote and developed areas. Each dot represents one part per million.

Source: *Adapted from U.S. Environmental Protection Agency, "Air Quality Criteria for Lead," (EPA-600/8-83-028B), 1984, p. 11-9.*

How Much Lead is Too Much?

The phrase "too much" lead can have two different meanings. The first meaning refers to the amount of lead in blood that has been shown to be associated with health problems (that is, the health-effect level). Lead above this point is "too much" because a person with that much lead has a good chance of experiencing some health damage.

The second meaning of "too much" refers to lead levels above some specified safety margin. Normally for most poisonous substances, a safety margin is set to provide some measure of security between an allowable or acceptable level and the level that is known to cause health problems. Since everyone does not react to every substance in exactly the same way, a safety margin provides some assurance that a very sensitive person will not show a health effect.

Safety margins also provide protection for a less sensitive person that comes into unexpected contact with the poison. For example, in the case of lead, if a child's blood level were below a safety margin, then a small, accidental or unexpected further exposure to lead probably would not raise the blood level enough to cause a health problem. However, if the amount of lead in the child was just below the health-effect level, that small extra exposure could easily push the level up enough to cause a health problem. A relatively frequent "extra" exposure for young children occurs when a small scrap of leaded paint is swallowed causing the blood lead level to go up by a large amount.

To determine the size of the safety margin for a specific pollutant, each substance must be evaluated separately. Two rough rules for setting allowable levels and safety margins are:

- for man-made substances, such as chemical pesticides, the allowable level is usually set at about 1% of the level that can cause health problems, thus providing a 99% safety margin;
- for naturally-occurring substances that do not cause cancer (are non-carcinogenic), an allowable level is usually set at about 10% to 40% of the health-effect level, thus providing a safety margin of 60% to 90%.

Lead is a naturally-occurring, non-carcinogenic substance. Applying the usual safety margin rule to lead would yield an allowable level of 2.5 to 10 μg/dl (that is, 10% to 40% of the 25 μg/dl health-effect level).

From the first perspective, the level where health problems are know to occur, "too much" (or an excessive blood-lead level) has been set at 25 μg/dl for children. For adults, there has been no actual level set, but many authorities feel that there should be concern for adults' health problems when the blood-lead level reaches about 30 μg/dl.

From the second perspective, the safety margin level, there is no clear agreement. Many government agencies and health departments in Canada (e.g, Toronto Department of Public Health, recommendation of the Royal

Society of Canada's Commission on Lead in the Environment) consider 20 µg/dl as an "intervention, allowable, or acceptable" level. This level provides only about a 20% safety margin below the health-effect level.

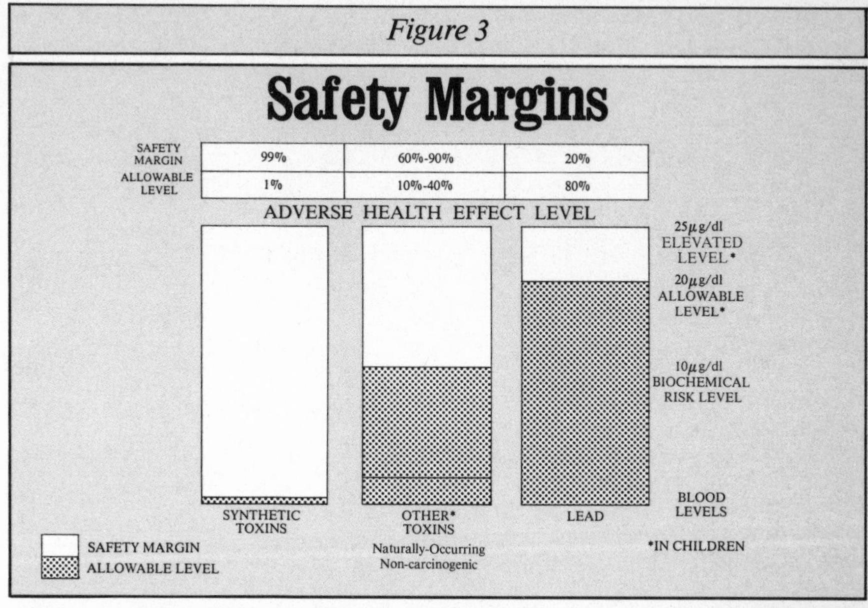

A comparison of the small safety margin for Lead with those for Synthetic Toxins and Other Toxins (naturally occurring, non-carcinogenic). A blood level of 10µg/dl for lead corresponds to the upper-limit, allowable level for Other Toxins. A blood level of 15µg/dl for lead would provide a 40% safety margin.
Source: *R. Elias, U.S. Environmental Protection Agency, personal communication, February, 1986; and J. Pirkle, Centers for Disease Control, personal communication, December, 1985.*

Many health professionals and scientists do not see the 20 µg/dl level as providing any safety margin at all. The basis for this view is that a number of changes are known to occur in children at levels below 20 µg/dl. Studies from a number of different laboratories have shown that beginning at about 10 to 15 µg/dl, fairly extensive biochemical changes begin to occur in the bodies of young children. These biochemical changes affect normal electrical activity in the brain, they cause a decrease in some important proteins in the blood that are involved in many different systems, and they can cause behavioural problems, such as hyperactivity. These internal biochemical changes are clear signs that if the lead level goes higher, more changes will occur in these internal systems resulting in the kinds of health problems that are evident at higher blood-lead levels.

Because of these widespread internal changes that begin to occur at a blood-lead level of about 10 to 15 μg/dl, this level is called a biochemical-risk level. If 10 μg/dl were set as the upper limit of safety for lead, then there would be a 60% safety margin below the agreed-upon adverse health-effect level of 25 μg/dl. Using this interpretation, 10 μg/dl appears to provide a good safety margin and would place lead in a comparable position to other, similar toxins.

However, at this point a problem arises. While 10 μg/dl is an adequate safety margin below the 25 μg/dl level, there are a number of changes that begin to occur in the body at 10 μg/dl that some authorities believe also should be considered as health effects in their own right. And to compound this problem, at the present time it appears that the degree of environmental contamination by lead is so extensive that it would be very difficult to bring blood-lead levels in the majority of children down below 10 μg/dl. As a further compounding factor, some biochemical changes are known to begin, in both children and adults, at blood-lead levels as low as 5 μg/dl.

Taking into account all of these problems as much as possible, "too much" (in the sense of health effects) will be considered in this book to be 25 μg/dl for children and 30 μg/dl for adults. In addition, we recommend an "allowable" level of no more than 15 μg/dl for children to provide some safety margin below the health-effect level. Note that "children" refers to ages six and under and adults, to ages 18 and over. The situation for ages 7 to 17 is somewhat ambiguous. In general, people in this age group tend to have lower blood-lead levels than young children or adults. Since the most lead-sensitive systems are fairly well developed by age seven, this group often is included in the "adult" category.

Is There Too Much Lead in Our Children?

The best available and most recent data on blood-lead levels in Canadian children comes from an extensive study carried out by the Ontario Ministries of Environment, Health, and Labour in 1984 on over 1200 children living in urban, suburban, and rural locations. The results of this study (detailed in Chapter 9) reveal that average lead levels were lowest in rural areas, higher in suburban areas, and highest in urban areas. Similar studies conducted from 1983 to 1985 by the Toronto Department of Public Health found levels above the average urban level for children living near a lead industry in an urban area. Figure 4 shows these levels graphically.

Although most children today will probably not develop such severe learning and behavioural problems as Carlos did, the Tri-ministry study provides a basis for estimating how many children have "too much" lead according to the 25 μg/dl health-effects level. These results can also be used to estimate the number of children over the 20 μg/dl "intervention" level and the 10 μg/dl "biochemical-risk" level. Children in any of these groups could develop some slight developmental deficiencies, with more severe problems in those over 20 μg/dl. As you may recall, Carlos' level was about 40 μg/dl.

Figure 4

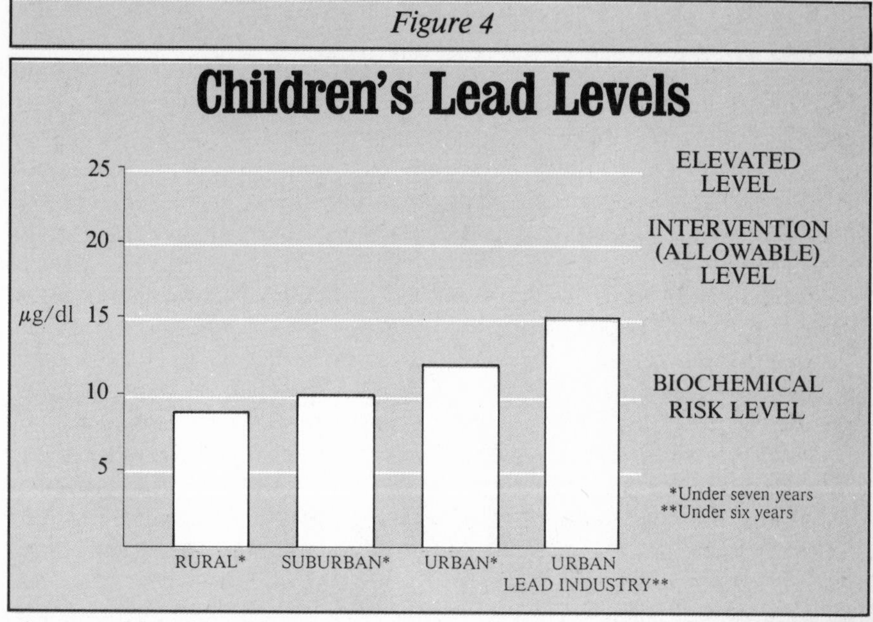

Average blood lead levels in μg/dl in young Canadian children in 1984 and 1985 for children living in rural, urban, and urban lead industry areas.

Sources: *C. Duncan et al, "Blood Lead and Associated Risk Factors in Ontario Children, 1984," Ontario Ministries of Environment, Health, and Labour, 1985, p. VIII-6; and E. Ellis, Toronto Department of Public Health, personal communication, May, 1986.*

The 1984 Tri-ministry study of blood-lead levels in Ontario found that 1.4% of the children under seven years of age had blood-lead levels over 25 μg/dl. There were 4.2% over 20 μg/dl and 51.8% over 10 μg/dl. If these percentages are applied to all children this age in Canada, it can be estimated that about 35,000 children may have "elevated" blood-lead levels, over 100,000 children may be above the "intervention" level, and almost 1.3 million may be at risk of minor developmental deficiencies due to biochemical changes. These numbers mean that a significant portion of our population of children may not realize its full potential in physical health, emotional maturity, and intellectual development.

Figure 5

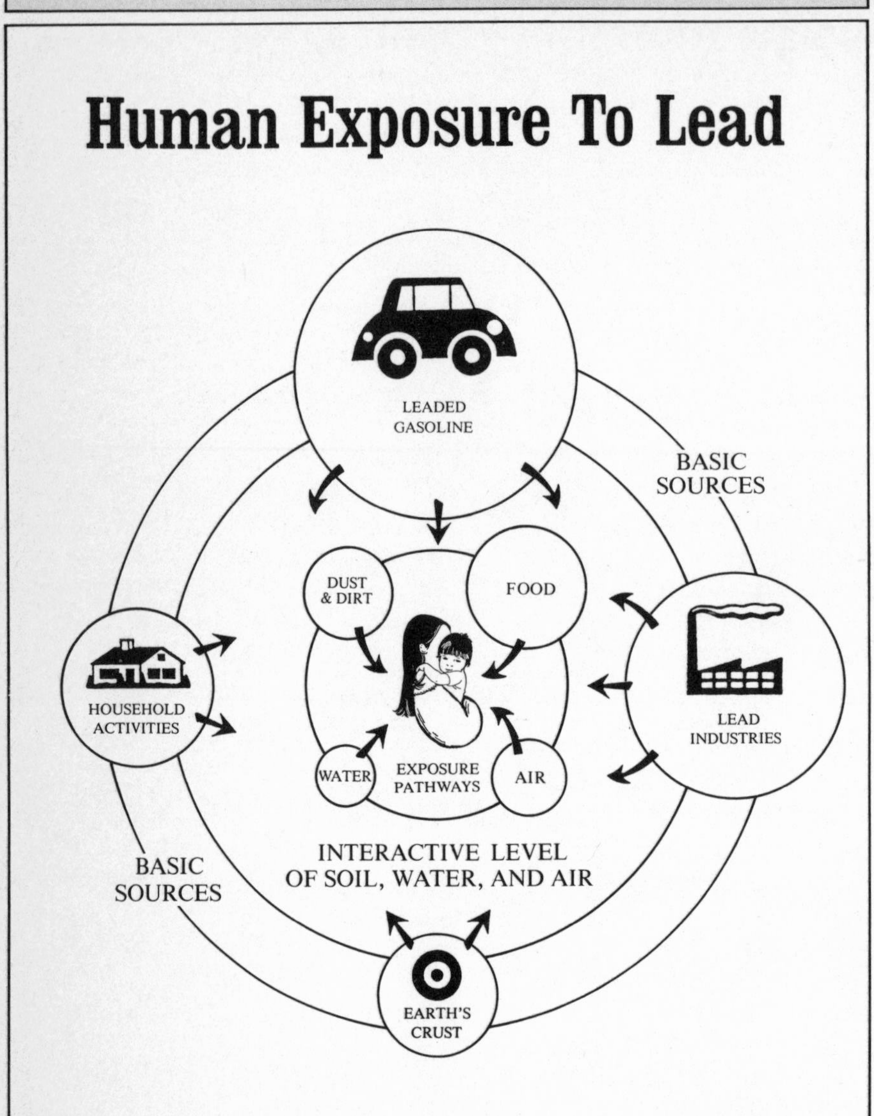

The four basic sources release lead into the environment. Environmental lead comes into direct human contact through food, air, dust and dirt, and drinking water. The size of the circles is a rough indication of the amount of lead in each source.

Source: *B. Wallace and K. Cooper,* ***Lead, People, and the Environment,*** *Niagara Neighbourhood Association, Toronto, Ontario, 1985.*

Where Is All the Lead Coming From?

Humans are exposed to lead from a number of sources, primarily in food, dust, air and water. Lead gets into these four sources indirectly from automobiles, industries, household activities, and the earth's crust. Other, more individualized sources of human contact with lead are occupational exposures, hobbies, smoking and alcohol consumption.

The amount of lead in many of these sources is subject to some degree of monitoring and control by the federal or provincial governments. However, there are serious problems with these attempts at control (see Chapters 12-14). Some important sources are hardly controlled at all. Of those that are controlled, most are in need of revision because they are based on out-of-date information. Many of the control strategies have not yet been developed in such a way that they can be legally enforced. And even the legally enforceable standards often do not succeed at real control, because occasional exceedances (e.g., air-lead levels above a legal standard) may not result in any changes that would keep such situations from occurring again. Although new additions of lead to the environment are much lower today than they were in the 1960s and 1970s, the persistency of lead is causing environmental levels to continue to increase, especially in urban areas or near lead industries. The next chapter describes those lead sources that cause the largest amount of exposure for most people.

Chapter 2

The Biggest Troublemakers

Most of the lead in our bodies comes from food or from dust that is breathed in or "eaten" from dusty hands, dishes, or foods. How lead gets into food and dust is a fascinating detective story. Only in the past few years have the major culprits (sources) been identified.

Finding the Evidence

The first, potential source to be examined was emissions from lead industries. Since blood-lead levels are higher in industrialized areas than in remote areas, it seemed possible that lead particles from the stacks of lead smelters and other lead industries were the major sources of lead in food and dust. However, when the amount of lead released by these industrial sources was measured, it became clear that some other sources were also affecting the amount of lead in the soil, food, and dust. There was more lead than could be accounted for by just the industrial sources. There had to be another contributor to the problem.

An important clue was found deeply buried in Greenland's polar icecap. Falling snow had "washed" dust particles, including lead particles, from the air. Most of the snow that fell in Greenland gradually became packed down into the icecap. Scientists were able to bore down into the ice to get samples of ice from known times in history. When these ice samples were analyzed, it became clear that there was a sharp increase in the lead content of ice formed in the 1940s. This fact became a critical clue that led to the ultimate identification of the culprit.

Automobile exhaust provided the next clue. This story begins in the 1920s when "horseless carriages" powered by gasoline began to be accepted by the general public. On December 9, 1921 in the General Motors Research Laboratory in Ohio, a far-reaching discovery was made — the addition of lead to gasoline resulted in a boost in power. The full story of lead as an additive to gasoline is told in Chapter 10.

Figure 6

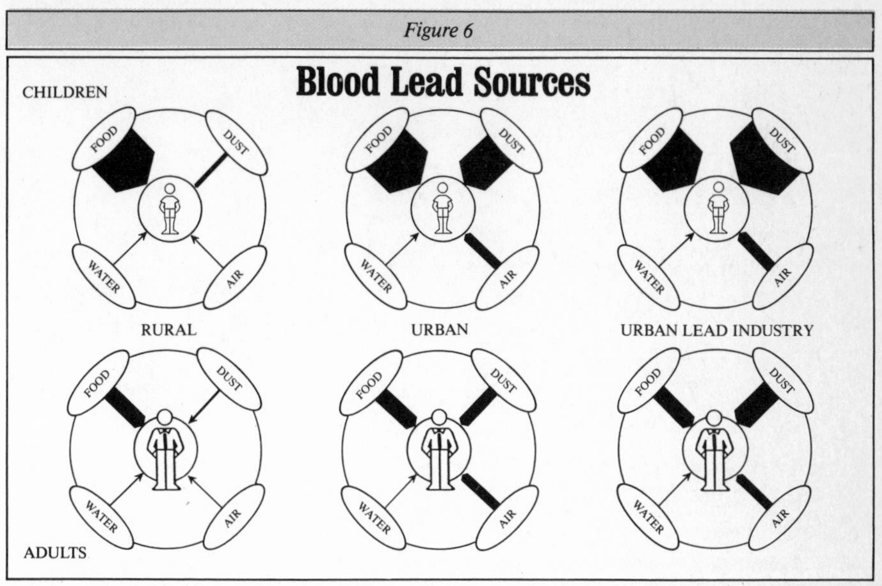

Relative size of the primary environmental sources that contribute to blood lead levels in two-year old children and adults living in rural, urban, and urban lead industry areas.
Source: *Appendix B.*

For many years there was no real recognition of the fact that lead added to gasoline might do anything besides improve engine performance. Some people expressed concern about possible risks to public health, but no studies were carried out to determine just what this risk might be. At first, leaded gasoline had little effect on the environment, because only one company sold it (they called it ''ethyl'' gas) and there were relatively few cars on the road. However, beginning in the 1940s, especially in the boom years after World War II ended in 1945, the situation changed. There was a rapid increase in the number of cars on the road. From then until about 1970, the majority of these cars were powered by leaded gasoline. Could this have been the cause of the increased lead levels found in the layers of polar ice in Greenland?

The critical answer to this question came from direct measurements on the amount of lead released by various sources: automobile exhausts, lead smelters, and others. These measurements revealed that the total amount of emissions from vehicles was much larger than from other sources. It turned out that about 75% of the lead in gasoline moved right through a car's system and came out the tailpipe. After the 1940s, lead was being spread so widely and in such large quantities through the environment that even the snow in Greenland contained significantly more lead than before.

Figure 7: Car exhausts from vehicles using leaded gasoline are the number one contributor to lead in blood.

Confirmation of the effect of automobile emissions on lead levels in the surrounding environment could be seen easily when the amount of lead in Toronto soil was measured in the mid-1970s. Figure 25 in Chapter 7 shows that soil-lead levels were higher along traffic corridors and in the congested downtown core than in other areas.

Some people, particularly those in lead-related industries, still did not believe that leaded gasoline contributed to the level of lead in blood. They wanted more direct evidence. The results of an ingenious experiment in the city of Turin in northern Italy provided just that kind of evidence.

The Turin experiment depended on the fact that lead from some mines around the world can be identified by unique characteristics (specifically the proportion of certain lead isotopes). For about three years, beginning in 1976, lead with an unusual characteristic was used in the gasoline that was sold in Turin. During these years, frequent measurements of lead in the air in Turin and in the blood of Turin residents were taken to see if that special kind of lead could be identified. The answer was clear. The new kind of lead was found in air and blood. Scientists were able to determine that about 24% of the lead in the blood of Turin residents was due to lead in gasoline. In a similar way, and taking into account lead from gasoline found in food, dust, and dirt, Environment Canada[1] has estimated that up to 40% of blood lead in urban children and 36% in urban adults is due to lead from gasoline. For children and adults living in rural areas, the comparable figure is about 15%.

Leaded Gasoline

The picture was becoming clear. Large amounts of lead were entering the environment from leaded gasoline, resulting in increased human exposure. Most of this gasoline lead drifted down on streets, yards, fields, and growing crops, thus adding to the lead in food and dust.

Further support for the role of leaded gasoline in human lead exposure came from a very large study conducted in the United States from 1976 to 1980.[2] One of a number of different health variables in this study was blood-lead level. The researchers noticed that blood-lead levels dropped significantly between the first and last years of their study just as sales of leaded gasoline were decreasing and those of unleaded gasoline were increasing. When the researchers plotted the yearly amounts of lead sold for gasoline use and the average blood-lead levels on a graph, the results were startling (see Figure 30 in Chapter 10). The amount of lead in gas and in blood went down, and occasionally up, simultaneously as if the two were walking along hand in hand. The evidence was strong. Gasoline lead was a primary determinant of lead in people's blood.

There seemed to be three ways in which lead in tailpipe exhausts could get into people's bodies. Some of the lead in the air could be breathed in directly. This pathway turned out to be of fairly minor importance. Some of the lead in the air settled into dust or dirt where it then could be picked up on shoes, pets, or playing children and brought into the house. A piece of food set on a slightly dusty surface and then eaten, or a slightly dirty finger going into a child's mouth were all that it took to bring the lead in the dust inside someone's body.

The pathway from lead in automobile exhaust to lead in food apparently also involved a simple process of lead particles settling out from the air directly on to food growing in the fields, food being transported to market, or soil in food-producing fields. Analysis of the lead content of different foods revealed that leafy vegetables often had lead on their surface while most of the lead in root vegetables was in the skin. Crops with easily washable skins that grew above ground, like tomatoes, beans, and squash, did not have much lead if they were well washed.

Food Processing

In the case of lead in food, it turned out that there is more to the story than just lead from gasoline. Analysis of lead levels in different foods revealed that there is more lead in processed foods than fresh foods. This meant that lead was being added during the processing, probably due to miscellaneous sources of lead in the equipment used to wash, cut, and prepare the food. Canned foods, particularly acidic foods such as juices and tomatoes, were found to have a much higher lead content than other processed foods. These lead increases were found to be due to lead in the solder used to form the side seam of the can. In the last few years, canned food processors have recognized this situation and have voluntarily begun changing their procedures to avoid the problem. Figure 10 in the next chapter will help you identify safe, lead-free cans.

Nevertheless, lead in gasoline turned out to be the biggest troublemaker for human exposure to lead. The obvious solution was simple: take lead out of gasoline. Putting this solution into practice has turned out to be a more difficult question primarily for economic reasons. Lead was originally put into gasoline to boost its octane level, which can also be achieved by other refining processes or the use of other additives. Removing lead and using other methods to boost octane required some adjustment in the gasoline industry. Some members of the industry have been reluctant to make these adjustments citing economic factors.

As with so many environmental questions, the choice seems to be between health and economics. Of course, we want better health, but can we afford the extra cost of unleaded gasoline? The truth is that we can have both health and financial savings!

The situation for a motorist is much more advantageous than for the industry. The major reason unleaded gas costs a little more is that it requires a slightly different and slightly more expensive type of refining process than leaded gasoline. However, all new cars and most older cars run much better on unleaded gas. For new cars, the slight price differential between unleaded and leaded gasoline is more than offset by better fuel economy and lower maintenance costs when unleaded gasoline is used (see Figure 8). For older cars, the difference in gasoline costs is about equal to the difference in maintenance costs and fuel economy.

In addition to a motorist's personal cost savings when he or she uses unleaded gasoline, there is an added economic benefit to him or her as a taxpayer. On the basis of cost-benefit studies in the United States, we estimate that the removal of lead from gasoline will save the Canadian government millions of dollars in reduced costs for health care and remedial education for persons damaged by lead.

It seems clear that discontinuing the use of lead in gasoline has relatively small costs and a number of significant benefits, including reduced environmental pollution, better health, and reduced costs for health care and remedial education. The time has come to get the lead out of gasoline.

Lead Industries

Everyone is exposed to lead from gasoline. City dwellers are exposed more than people in the country simply because there are more cars in the city.

Figure 8

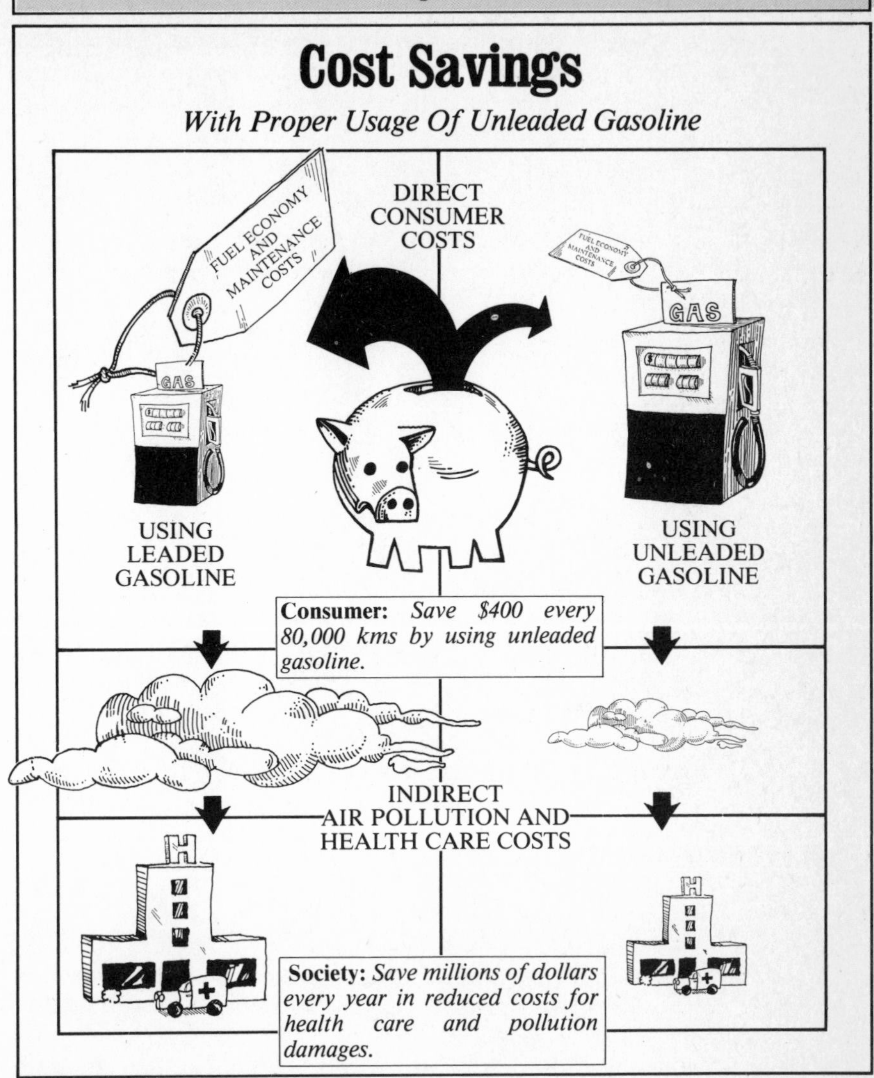

A comparison of direct consumer operating costs and indirect social costs of misfueling (i.e., using leaded gasoline in cars designed to use unleaded gasoline). Motorists who misfuel lose about $400 every 80,000 kilometres. Although misfuelers have a small cost savings at the pump, this savings is more than offset by the resulting poorer fuel economy and greater maintenance costs. Misfueling also costs society millions of dollars (i.e., more taxes) for increased costs of health care and air pollution damages.

But lead in gasoline is not the only source. Some people, like Lila Jarrett and her family, experience additional exposure because of where they live.

Lila Jarrett found out that living near a lead industry can cause special problems. Lila is like most of us. She likes a reasonably clean, but not spotless, home. However, when she and Joel married and moved into their first place, an older house in a mixed industrial-residential neighbourhood, she could not keep up with the dirt. It seemed that no matter how often she cleaned and dusted, within a short time there was so much dust she could write her name in it. After a few months and part-way through her first pregnancy, Lila found out that everyone in her neighbourhood had a dust problem. She also found out that there was a lead smelter behind her back fence and that lead, including the lead in her house dust, could be a problem for the baby she was carrying and for her child after it was born.

Lila began cleaning even more frequently and made other changes to reduce the dust level in her house. Everything helped a little bit, but apparently not enough. Lila wanted to protect her child from lead effects as much as possible. When Sara was only a few months old, Lila had her blood tested for lead. Sara had 16 μg/dl in her blood and she wasn't even old enough yet to be crawling around on the floor. The doctor told her that Sara did not have any lead health problems, but Lila found out that some biochemical changes could be occurring at 16 μg/dl and that Sara's level probably would go up once she began playing in the yard. Lila and Joel were not willing to take that kind of risk. For Sara's sake, they just moved away.

For people who live near poorly-controlled lead industries or who work with lead, these exposures can add significantly to the amount of lead in their bodies and can force changes in the way they wish to live. For them, the biggest troublemakers are both industrial exposures and lead in gasoline.

Chapter 3

Reducing Your Personal Risks

The bad news is that lead has the potential to cause undesirable, serious, and sometimes even irreversible effects on the quality of life of almost anyone. The good news is that many of these problems are preventable. Each of us can make choices that will reduce the risk of lead problems in our life. The situation in which we find ourselves at present is simply the result of our past choices. Although it is often easier to do things in old, habitual ways, it is both possible and wise to act differently in order to reduce the risk of health problems due to lead. This chapter describes some of the ways you can reduce lead risks for yourself and your family.

Lead does not cause human health problems until it is absorbed into the body. Absorbed lead is what is measured by blood-lead tests. Some previously absorbed lead is stored in bone or soft tissues and therefore, would not be measured by a blood-lead test.

The path from environmental lead to health effects has several steps. The first of these, *exposure,* refers to a person's contact with lead in food, dust, air, water, and a few other miscellaneous sources. The second step, *intake,* refers to that amount of the lead that is actually taken into the lungs or digestive system. The third step *absorption,* refers to that fraction of the lead taken into the body that is absorbed into the blood. And the final step, *health effects,* refers to mild or severe health problems that develop from sufficiently high levels of absorbed lead in the blood.

A continued daily *absorption* of 250 μg can cause severe lead poisoning. A *blood-lead level* of about 100 μg/dl can cause death. The difference between the two numbers lies in the fact that the absorption figure represents new lead coming into the body on a daily basis, while the blood-lead figure represents a body burden per decilitre of blood.

There are actions an individual can take to reduce the risk of health effects at each of the first three steps. If you can reduce your exposure, there is less intake and less absorption to worry about. Since exposure can never be reduced to zero, some of the suggestions are designed to reduce intake and absorption as well. The following sections describe risk reduction actions for:

- everyone;
- young children;
- gardeners or farmers;
- renovators or hobbyists; and
- people who live near a lead industry.

Risk Reducers for Everyone

Although anyone can be damaged by too much lead, there are certain groups of people who are more sensitive to lead effects or who tend to have greater exposure to lead than others. The most sensitive group is children under the age of about six years, with the most sensitive sub-group being those two and three years old. Another very sensitive group is women of child-bearing age. The women themselves are not at risk, but if they are pregnant or ever plan on becoming pregnant, their unborn child could be at risk. A fetus is extremely sensitive to lead, and lead in the mother's blood (which could come from past lead exposures stored at various sites in the mother's body) easily passes directly to the fetus.

Another group at greater than usual risk is anyone who is occupationally exposed to lead, either directly in a lead industry or indirectly in another industry that uses lead. Families of such workers may also have increased exposure if the worker does not change clothes or, when appropriate, shower before coming home. Adults who handle leaded materials (such as solder or paint) in hobbies or renovation activities may also experience quite high lead exposures, even from a single day's use. Lifestyle choices for all these groups are covered in the following sections.

Diet

Since for most people the largest lead exposure source is food, diet is a very important consideration. Eating fresh, frozen, or dried food rather than processed (especially canned) foods is one of the most important dietary changes that can be made to reduce lead intake. When fresh vegetables are used, their lead content can be reduced by careful washing of leafy vegetables and peeling of root vegetables (see Figure 9).

Processed foods in general have slightly larger amounts of lead than do fresh foods simply because of the greater opportunity for contact with lead or leaded substances in cutting, handling, mixing, and other food processing procedures. If the processed food is then placed in cans with a lead-soldered side seam, it is possible for the lead level in the food to increase by a large amount. Up to ten times more lead than in fresh foods has been measured in canned acidic foods or beverages, like tomatoes or juices.[1]

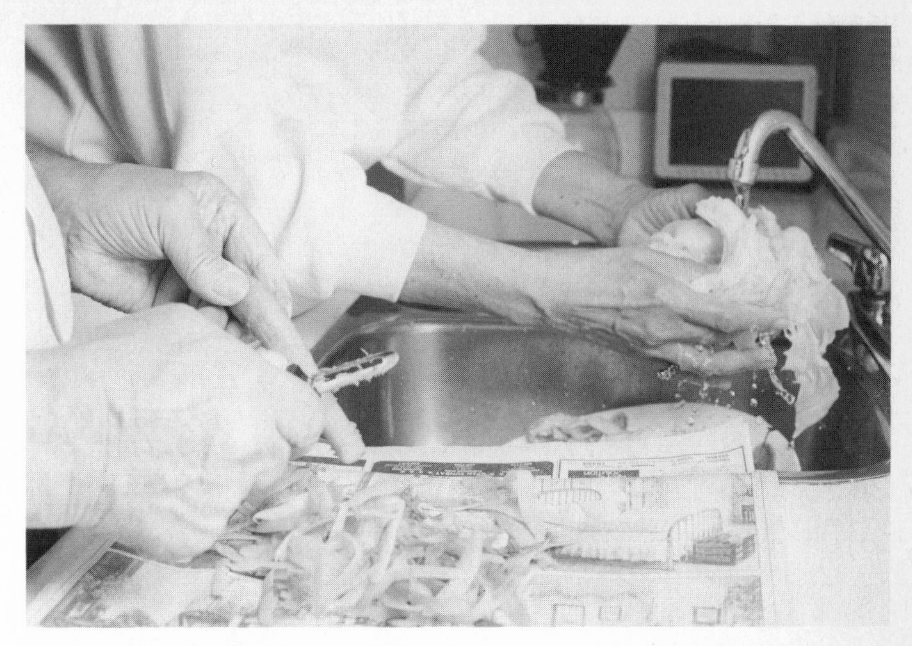

Figure 9: Careful washing of leafy vegetables and peeling of root vegetables can reduce your exposure to lead in food.

If foods are left in an open lead-soldered can for any length of time, the lead level may increase by a large amount, (e.g., up to eight times over a five-day period).[2] There have been reports of children becoming lead poisoned by drinking fruit juice that had been stored in an open can.[3] Dented or damaged cans also may have a higher lead content than intact cans. If possible, purchase juices and tomato products in plastic or glass containers or lead-free cans rather than in lead-soldered cans. See Figure 10 for information on identifying lead-free cans.

As was discussed earlier, the last step in the pathway of lead from the environment to health effects is absorption of the lead from the lungs and digestive system into the bloodstream. There is not much you can do to change the amount of lead absorbed in the lungs, but there are simple dietary habits that can reduce the amount absorbed from the digestive system.

Children absorb into the bloodstream about half (50%) of the lead in their digestive system. Adults absorb only about 10%. Either group may absorb higher percentages at times when there is little food in the digestive system. There is also greater absorption in people who have low levels of calcium, iron, and zinc in their systems. Eating adequate amounts of foods that are high in these three substances will decrease your absorption of lead thus providing a simple way to protect your health. Most adult diets probably have adequate amounts of zinc, but children may need to ensure an adequate zinc supply by eating meats or seafood. Many adults and children

Figure 10

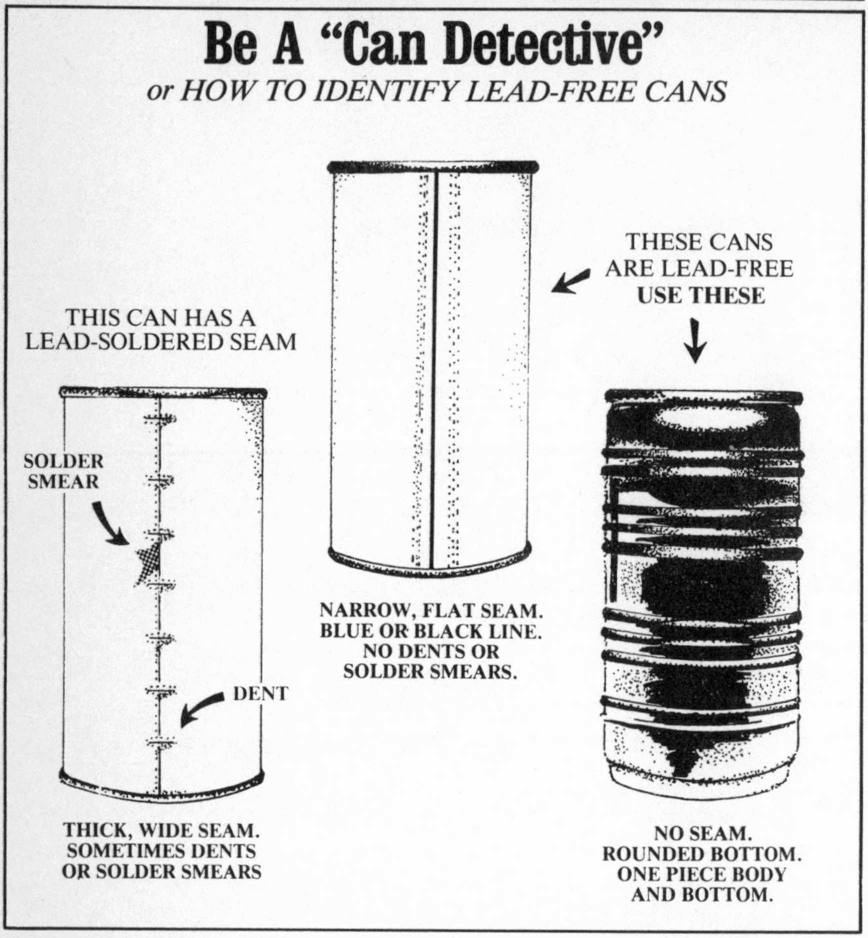

Be A "Can Detective"

or HOW TO IDENTIFY LEAD-FREE CANS

THESE CANS
ARE LEAD-FREE
USE THESE

THIS CAN HAS A
LEAD-SOLDERED SEAM

SOLDER
SMEAR

DENT

NARROW, FLAT SEAM.
BLUE OR BLACK LINE.
NO DENTS OR
SOLDER SMEARS.

THICK, WIDE SEAM.
SOMETIMES DENTS
OR SOLDER SMEARS

NO SEAM.
ROUNDED BOTTOM.
ONE PIECE BODY
AND BOTTOM.

Step 1. *Is there a side seam or is the can seamless with a rounded bottom?*
If it's seamless with a rounded bottom, it's LEAD FREE.
Step. 2. *If there is a side seam, does it have a thin blue or black line running along the seam?*
If there is a thin blue or black line, it's LEAD FREE.
In addition to these two big clues, the seams on LEAD-FREE cans are usually flatter and narrower than on lead-soldered cans, and they do not have "dents" or solder "smears." See drawings.
Source: *Consumers Union Foundation, "Action Kit: Lead Hazards and Children," Institute for Consumer Policy Research, Mount Vernon, NY, no date.*

have diets that are low in calcium and iron. Food sources rich in these two substances are shown in Table 3.1.

Table 3.1

Sources of Calcium and Iron

Calcium — excellent sources: beans, cauliflower, chard, cheese, cream, egg yolk, kale, milk, molasses, rhubarb.

good sources: almonds, beets, bran, cabbage, carrots, celery, chocolate, dates, figs, kohlrabi, lemons, lettuce, oatmeal, oranges, oysters, parsnips, pineapples, raspberries, rutabagas, shell fish, spinach, turnips, walnuts, watercress.

Iron — excellent sources: almonds, asparagus, bran, beans, Boston brown bread, cauliflower, celery, chard, dandelions, egg yolk, graham bread, kidney, lettuce, liver, oatmeal, oysters, soy beans, whole wheat.

good sources: apricots, beets, beef, cabbage, cornmeal, cucumbers, currants, dates, duck, goose, greens, lamb, molasses, mushrooms, oranges, parsnips, peanuts, peas, pineapple, potatoes, prunes, radishes, raisins, rhubarb, tomatoes, turnips.

There is less iron in carrots and milk than in other foods. Only 50% of the iron in spinach and similar vegetables is assimilable by the body.

Source: *L. Thomas,* ***Taber's Cyclopedic Medical Dictionary,*** *12th ed., F. A. Davis Co., Philadelphia, 1973.*

Other helpful guidelines for reducing lead problems are shown in Table 3.2, "Diet and Food Preparation Guidelines." Since we are exposed to much more lead than people were fifty years ago, all of us can benefit from following these guidelines. However, careful attention to these guidelines is especially important for children, women of child-bearing age, people who are occupationally exposed, and others with occasionally high exposure levels.

Other "dietary" sources of lead are bonemeal and dolomite, which some people take to increase calcium in their diet. Adequate calcium levels are very important in reducing lead absorption, but bonemeal and, in some cases dolomite, should not be used for this purpose since they may contain considerable amounts of lead. Bonemeal, in particular, may be high in lead because it is made from animal bones and animals, like humans, store lead in their bones. Dolomite, a calcium-bearing rock, may contain some lead depending on the natural lead level in the area where the dolomite originated.

Table 3.2

Diet and Food Preparation Guidelines

- Eat more calcium, iron, and zinc.
- Do not use canned food or beverages except those in lead-free cans.
- Eat an adequate amount of protein, neither too much nor too little.
- Peel all root crops.
- Wash leafy crops well and discard outer leaves.
- Eat young plants since their lead level is less than older plants.
- Although not as well confirmed as the above guidelines, it has been suggested that cutting down on caffeine, fats, and fatty foods may also reduce lead absorption.

Smoking cigarettes and drinking alcoholic beverages, especially wine, also can increase the amount of lead in your blood. Alcoholic beverages apparently contain lead because of fermenting, aging, and sealing processes. Because of processing differences, there are wide variations, ranging up to 300 μg per litre, in the amount of lead in the finished product.[4] Each cigarette contains from two to twelve μg of lead. About six percent of this lead is inhaled with most of the rest ending up in house dust.[5]

Recent research has shown that, compared to non-smokers and non-drinkers, female smokers have blood-lead increases of one to two μg/dl, male smokers two to three μg/dl, and both male and female drinkers about one μg/dl. For females that both drink and smoke, the total increase is 2.5 to 3 μg/dl and for males, 2.5 to 3.5 μg/dl.[6] Blood-lead increases of these sizes, combined with normal exposure levels plus other factors such as occupational exposure or renovation activities, may be enough to cause health problems.

Lifestyle Changes

Switching to unleaded gasoline, if you have not already done so, will reduce your lead exposure in two ways. Filling a gas tank with leaded gas, can expose you to the mostly easily absorbed source of lead — inhalable fumes of organic lead that can go into your lungs and from there into your brain. Lead in gasoline also can be absorbed directly through the skin. In addition, using leaded gas releases lead from the exhaust system into the environment where it can come back to you, as well as to everyone else, through food and dust. The first problem can be reduced by taking special care to avoid breathing the fumes when putting leaded gasoline in your automobile tank. Both the first and the second can be eliminated by switching to unleaded gasoline.

Figure 11: Some things to avoid if you want to reduce your lead exposure.

Figure 12: Some things to eat if you want to reduce your lead absorption.

Table 3.3

Adult Risk Reducers

- Follow the food guidelines in Table 3.2
- Stop smoking.
- Reduce alcohol consumption.
- Use unleaded gasoline.
- Wash your hands before eating.
- Watch out for lead exposures in hobby activities.
- Be a dustbuster.
- Check soil-lead levels in city gardens.
- Protect yourself and your family when soldering or renovating.

Some simple acts of personal cleanliness may reduce the lead exposure for yourself and for others in your home. Always washing your hands before touching food or dishes is a lead-reducing habit to establish. Keeping clean fingernails is important too.

If you have a hobby that uses lead, reduce your risks by adopting some protection guidelines. First, make sure that you protect yourself from unnecessary exposure. If appropriate, wear a mask or gloves. Because of its skin absorption characteristic and easily inhaled fumes, do not use leaded gasoline as a cleaner or solvent. Investigate the possibility of choosing alternative materials that do not contain lead. Second, make sure that others in your home are protected. Keep children away while you are working. Put any leaded or other hazardous materials in a place where other people who may not know what they are have no access. Third, do not let your activities increase the lead level in the household. Arrange ventilation so that lead in the air is exhausted to the exterior. Carefully contain and clean up any particles of lead. Keep your hobby clothes separate from the family laundry. Your hobby will be more enjoyable to you if you know that it is not causing lead problems for yourself or others.

Clean House

Dust and dirt in urban areas, near traffic routes, or near industries may contain relatively large amounts of lead. The best way to avoid lead exposure problems from dust and dirt is to keep a clean house. Several ideas for reducing dust levels inside the house are summarized in the "Dustbusters" table.

If you have reason to think the level of lead in your house dust may be high, you might wish to do a thorough "decontamination" cleaning and then follow it regularly with the dustbusting routine. "Decontamination" cleaning involves wet cleaning with a phosphate detergent or cleaner

Table 3.4

Dustbusters

- Use a damp mop and damp cloths for cleaning and dusting.
- Sweep only when absolutely necessary and in a careful manner.
- Clean near windows and doors frequently since dust in these areas tends to have a high lead level.
- Use window shades that can be wiped clean and washable curtains.
- Vacuum or wash window screens periodically.
- Vacuum rugs often. Cover them with a sheet when babies play on them. If your vacuum expels dust, don't vacuum when children are around. Flat weave carpets and bare floors collect less dust than shag, pile, or looped rugs.
- Have a professional clean your forced-air heating ducts periodically.
- During the heating season, clean or replace furnace filters every month.
- Try to keep the humidity level between about 35-50%. Drier air encourages dust to fly around and wetter air encourages mould growth.
- Brush house pets often. Their coats collect dust and dirt. Groom them outside and keep them out of young children's rooms.
- Clean children's bedrooms or other play areas often. Remove dust catching items or store them in closets or drawers. A covered toy box keeps toys cleaner and makes cleaning easier.
- Cover furniture with washable covers.
- Clean shoes or remove them when you come inside.

Make Dust Control Your Goal

Source: *Adapted from "Dustbusters: Tips for controlling lead exposure through housedust," no date. Used with the permission of the Toronto Department of Public Health. Readers may reproduce this table provided the Toronto Department of Public Health is acknowledged.*

(phosphate[7] seems to help loosen lead particles from surfaces) in every nook and cranny of your house. Make sure that no children are exposed to the dust that gets disturbed or to the dirty water. Change the cleaning water frequently and/or go over the surfaces a second time with clean water. Carpets should be cleaned twice using a commercial carpet cleaning machine: follow the directions using a hot water-phosphate-detergent mixture, wait 24 hours, and then repeat the entire cleaning process using the regular commercial cleaning solution. If the carpet is removable, do both sides and the floor underneath.

You know your home and situation best. Look around and see where changes might be made that would reduce lead dust. If your entry way tends to get dirty easily, perhaps you should consider a change from outside to inside footwear at that point. Maybe you should use washable door mats. Note that these should be kept separate from the regular laundry. Is it possible to hose down the exterior of the entrance, walkway, or driveway? Is dust more likely to come in some windows than others and can the "dustier" windows be opened less frequently? Is there a worker in the family who may bring in clothes or shoes that have been exposed to lead on the job? If so, these clothes should be kept separate from the regular laundry.

If you have a fireplace, wood-burning stove or furnace, or an outdoor fire location, avoid burning materials that might contain lead, such as painted wood, used oil, old batteries, coloured printed matter (e.g., magazines, newspapers, advertisements), printed food wrappers, scrap metal, and building scraps. Burning lead-containing substances can release the lead in a form that is quickly and easily absorbed resulting in a large increase in exposure.

De-leading Your Child's Environment

Since children under six years of age have much greater exposure to lead and much more serious problems from such exposure than do older children or adults, it is important to keep their exposure to lead as low as possible. Every place where a child spends time is a potential source of lead exposure. Day care centres, nursery schools, gym classes, or any other location with activities for pre-school children should develop guidelines for protecting children from lead.

The best way to decide what changes should be made in a particular situation is to observe carefully where the children spend their time and what they do there. Especially focus on play and eating behaviours. In addition, examine their diet to see if some changes could be made there to reduce their exposure to or absorption of lead in food.

Consider where and how your children eat their meals and between meal snacks. Make sure that any surfaces that the food touches have had the dust wiped off recently. (Dust refers not only to a visible layer of dust, but also to the invisible layer that constantly settles on everything.) If you are having a picnic, eat on a recently cleaned table or a cloth rather than on the ground and try to keep all dirt away from the food.

Since one of the largest sources of lead for most children is dust and dirt on their hands or under their fingernails, make sure their hands are washed well before they eat a meal or snack, and that their fingernails are cleaned frequently. If they are eating while playing or while out shopping or on a trip, wash their hands first and do not let them eat food that has been set down on some questionably dirty or dusty surface. If food drops on the floor or the ground, either discard it or wash it well before giving it back to the child. Also be sure that the adult or older child who prepares or serves food or drinks has recently washed his or her hands and used clean utensils.

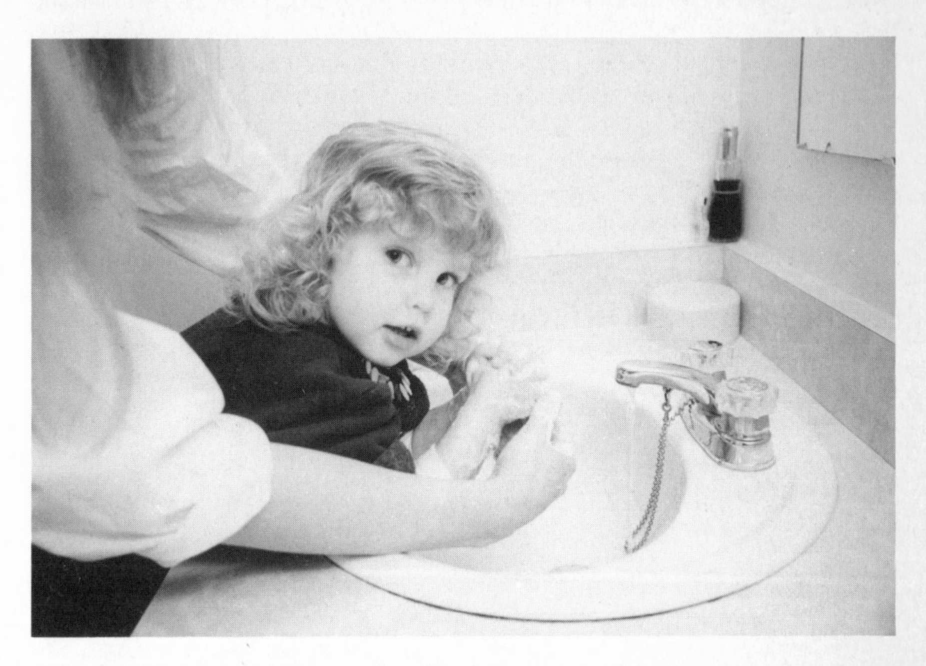

Figure 13: Careful and frequent handwashing is one of the best ways to reduce a child's lead exposure.

Examine outdoor play areas, especially areas under swings or at the end of slides. If there is bare soil, change the situation by planting hardy grass or by covering the area with clean sand, wood chips, old rugs, or mats. For play in earth or sand, provide a play box that has been filled with clean earth or sand. Look and see how close the play area is to moving vehicles or parking spaces. Within the first 10 to 20 metres of high-traffic areas, soil-lead levels may be considerably higher than at greater distances.

Indoor play areas should be easy to clean and as free of dust as possible. For the most sensitive two and three year olds, daily vacuuming or wet mopping, and wet dusting of their primary indoor play space may be desirable. Pay special attention to areas around windows and doors.

Choose toys with surfaces that are easy to clean. A plush doll may act like a duster. Since a child may take in quite a bit of lead dust by sucking or mouthing a fabric or plush toy, these types of toys should be washed frequently. Painted toys may present an undesirable degree of exposure. Paint used on children's furniture and toys may be marked as safe for children. However, the amount of lead allowed in all paints (interior, exterior, and for children's articles) can be as high as 5000 parts per million

Figure 14: Bare soil in a children's play area should be resodded or covered with sand or mats.

Figure 15: The best place for women of child-bearing age and young children is somewhere else when major home renovations or paint removal is occurring.

(as of March, 1986). Many health experts feel this level is too high for articles for young children. Young children should not be allowed to play with toys with peeling and chipped paint. Children who chew on toys should not be given painted toys until they are over that stage.

Also avoid unusual "toys" such as matches, scrap metal, or some adult work materials that might contain lead. Burning materials that contain lead can release lead fumes that can be inhaled very easily. For this reason, as well as for other safety reasons, do not let your child burn things. In addition, avoid smoke or fumes from other types of fires unless you know that no objects containing lead are being burned.

Table 3.5

Risk Reducers for Pre-school Children

- Follow the food guidelines in Table 3.2.
- Wash hands frequently and always before eating.
- Keep fingernails clean.
- Wet mop and wet dust play areas several times a week.
- Keep a clean house.
- Cover any bare-soil play areas.
- Do not let a child chew or suck painted toys or furniture.
- Re-do areas with peeling or chipped paint following the renovators' guidelines.
- Go on an outing while renovations are being made.

Hobbies and renovation activities may be the most frequent source of additonal lead exposures to both children and adults. If renovations, soldering, stripping of painted furniture, stained glass work, or some hobby activities involving lead are occurring in your home, make sure that these activities do not cause an increase in lead exposure to children. Old painted surfaces may have very high levels of lead. Even recently painted surfaces can have a lead level that is risky for young children. A single small chip of leaded paint may have ten times more lead than a child's usual total daily diet. Keep children away from renovation, soldering, or similar lead-related activities and be very careful to clean up any scraps or debris from this kind of work.

Most children suck their fingers or chew on toys during some period in their life. If your children are in this stage, wash their hands and toys frequently, keep their nails clean, and try to discourage these habits. Careful attention to their behaviour at such times may give you clues for things you can do to reduce such activities. For example, absent-minded finger sucking may be dropped if someone begins a simple hand-playing game with the child. Or, accepting a recently cleaned toy may keep a child from mouthing a dirty toy.

Some children go through a stage in which they eat non-food items such as erasers, paper, dirt, etc. This kind of eating is called pica (pronounced pie-ka). Pica can result in quite high and serious lead levels. For example, just chewing a piece of brightly-coloured printed matter may cause a significant increase in lead intake. Since concentrated lead has a sweet taste, any lead-containing objects that a child chances to mouth or eat may become a preferred item. Some children seem to become quite fond of lead paint chips. This kind of behaviour should be strongly discouraged. It is thought that some cases of pica occur because of emotional stress or because the child is missing some essential substance in his or her normal diet and is trying to find a substitute. If your child has pica, you might wish to discuss the situation with your doctor. Also make sure your child has a balanced diet.

Some of the suggestions described in this section for reducing your child's lead exposures may seem difficult to achieve. If you can't do all of them, your children will benefit from as many as you are able to do. Start by making the changes that are easiest for you to do and gradually add other changes. Remember that an extremely high level of lead protection is only needed for the first six years of life and is most important for two and three year olds. It may help you to make these changes if you keep in mind that by doing them you are increasing your children's chances for reaching full intellectual potential, emotional stability, and quality of health for the rest of their lives.

Gardening and Farming

If you live in an area that is remote from industries, old lead works, or traffic or one that has never had sewage sludge applied as a fertilizer,[8] your soil will probably not have a high lead level. Almost anyone else, especially city gardeners, may find that their soil has more lead than is desirable for growing food. Find out by getting your soil tested. Most areas have facilities available to the public for testing soil for heavy metals. (Appendix I contains soil-testing contact information.) Often you can ask for tests for cadmium, zinc, and possibly copper, at the same time that a test for lead is made. The *Gardening Guide* (Figure 16) shows crops that are safe to plant in soils at various lead levels.

If your soil is high in organic matter and at an approximately neutral acidic level (i.e., a pH of about 6.5 to 7), most of the lead that is present in the soil will become bound to soil particles in a way that prevents it from being incorporated into growing crops. You can adjust your pH level if it is too acidic (under 6.5) by adding wood ashes (but not ashes from painted wood which may have contained lead) or an appropriate commercial additive. Organic matter can be added by using kitchen scraps that have been composted. Do not use sewage sludge as fertilizer unless the actual batch

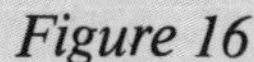

Figure 16

SOIL LEAD LEVEL IN PPM	KINDS OF VEGETABLES ROOTS	LEAFY	FRUITING
0-500	OK	OK	OK
500-1000	NO	OK-NO*	OK
over 1000	NO	NO	OK

**OK if garden is not near a main traffic route or smelter, and NO if it is.*

Foods that are safe to grow (and eat) in soils containing different amounts of lead. OK means safe; NO means not safe.

Source: *Adapted from City of Toronto Department of Public Health, "Get the Lead Out," no date.*

of sludge you receive has been tested and found to have acceptable levels of heavy metals.[9]

Some areas are not desirable for growing food crops. They include areas around smelters, incinerators, highways, snow dump sites, landfilled areas, land lying within ten to twenty metres of a heavily travelled road, and land near gutter downspouts or near painted buildings.

Renovations and Hobbies

Frank was proud. It was his first apartment and he had finally gotten it down to the bare framework so he could begin decorating it the way he wanted it. He knew it was going to look really good when he finished with it. Now he could work with new materials instead of just taking down the old stuff. When he started, he had never thought that it would take him a full month to take down all the old ceilings and remove the broken plaster and other piles of scrap and dirt. What a dirty job! Some days it was enough to make him really choke. But Frank had been careful about protecting himself from breathing in all that old dust. On the dirtiest days he had worn a disposable dust mask despite the fact that the mask was hot and made it harder to breathe.

A few days after Frank finished the first stage of his renovations, he noticed a group of people right in front of his house. They seemed to be waiting in line at a health department van. He saw one of his new neighbours in the line and asked him, "What is all this about? What's happening?"

"Well, they're testing our blood again."

"Why are they testing blood?"

"Most years they just check the kids' blood, but this year they're testing the adults, too. It's for lead, you know."

"But why are they testing for lead? And why are they doing it here?" asked Frank.

"Don't you know there is a lead smelter right behind our houses? We've got more lead in our blood than others. They came down here just to make sure they test everybody."

"What do you mean a lead smelter? Nobody told me anything about it before. Is it dangerous to live here? Should I get my blood tested too?"

"Well, they say, if you live here, get your blood tested. So you'd better get in line."

Frank joined the others and got his blood tested. Sometime later he received a letter from the health department telling him that his level was a little high, just over 20 $\mu g/dl$, and that he should come in for another test. A month after his original blood test, Frank came in for his retest. He was upset and asked the doctor, "Why was my level so high? Is it going to cause problems for me?"

The doctor told him, "There is no need to be concerned right now. Let's wait until we get the results from this new test. Now, tell me. What kind of work do you do? Do you have any hobbies that use lead? Have you been doing any renovations?"

Frank said, "I work in an office and my hobbies are swimming and running. But I was doing a lot of renovations just before my first test. How could that affect my blood?"

"If you were working with any material that had old paint on it, you could have been exposed to lead dust. Or you may have disturbed old piles of dust behind panelling or in cracks and crevices. That dust may have had a fairly high level of lead because you live near the smelter."

Frank was outraged. "Why didn't somebody tell me that it was a health risk to renovate an apartment in that area? I wore a mask to keep from breathing in the dust. Why didn't it keep the lead out? What about other people that move in and don't know?"

The doctor agreed that lack of knowledge about the danger of lead is a big problem and explained that small disposable masks do not provide enough protection. Fortunately, Frank's slightly elevated level dropped fairly rapidly after he finished the first stage of his renovations. (See chapter 8 for what happens to lead inside the body.) The results of the second test showed a blood-lead level of only 12 μg/dl. Frank, however, remained concerned about these issues and thought that someone, either the city or his landlord, should have told him about the health risks associated with renovations, especially renovations in an area near a lead industry. He joined the neighbourhood lead committee and began working on ways to improve the situation in his area.

Doing home renovations is a good way to personalize your apartment or home. But as Frank found out, it can increase your exposure to lead. Even in areas not near a lead industry, renovations can cause problems because of the high lead levels in older paints. A 1981 survey of Washington, D.C. families who had recently renovated their homes found high blood-lead levels averaging 47 μg/dl in the children and 41 μg/dl in the adults.[10]

Until about a decade ago, most house paint, both interior and exterior, contained lead — sometimes at levels as high as 500,000 parts per million (ppm). Current levels of lead in paint in Canada, even in paints labelled "lead-free," can go as high as 5000 ppm. To determine if lead is present in your paint, see if the paint (including all underlying, hidden layers) turns black when a 10% solution of sodium sulfide is applied to it.[11]

Sanding leaded paint off an old door frame to reveal the wood underneath may cause very large increases in lead in the air and dust. Burning off such paint produces volatized lead fumes that can be inhaled easily. Sawing through, or knocking down, painted walls and many other typical renovation activities can cause fresh releases of lead to the environment or may open up old caches of lead dust that had settled into cracks or behind boards.

Figure 17: Appropriate protection gear and clean-up arrangements for removing leaded paint. Note the dual-filter mask, which is available at Safety Supply distributors nationwide.

Table 3.6

Renovators' Guidelines

Soldering

- Provide for ventilation with the air moving away from you and from the rest of the house.
- Wear a dual-filter mask such as the one in Figure 17.
- Clean up any tiny scraps of solder carefully and dispose of in a closed container in the garbage.

Removing Old Lead Paint

- Remove or cover with plastic any carpets, rugs, upholstered furniture, clothing, cooking utensils, etc.
- Put drop cloths in the working area to catch any paint chips.
- If possible, do only one room at a time, remove furnishings, and close off the room.
- Do not use a power sander since it can create too much lead dust.
- Wear a dual-filter mask (see Figure 17), coveralls, hat, and shoe covers.
- During work, close the windows to prevent lead dust from entering the surrounding environment and keep children, women of child-bearing age, pets, and adults with heart or lung conditions away until the clean-up has been completed.
- Do not eat or smoke in the work area.
- Use a heat gun to heat the painted surface; then scrape with a sharp object. With a heat gun, an exhaust ventilation fan (not an air conditioner) is needed.
- At the end of each working session, put the chips in a double plastic bag, close it, and put it into the garbage.
- Vacuum and wet scrub the area with a high phosphate detergent such as Spic and Span. Do not sweep or dry dust.
- Remove your protective clothing and shoe covers in the work area to avoid carrying lead dust into the house.
- Shower and shampoo at the end of each work session.
- Wash your work clothes separately and wipe your footwear off with water and a rag that is washed with the work clothes; or, if you are doing a large job, wear clothing and shoe covers that can be sealed in plastic bags and thrown away.
- Wash your hands thoroughly and rinse out your mouth before eating or smoking.
- Soon after removal of the old paint, repaint the surface. Avoid spray painting if possible; if you paint frequently, wear a mask while brushing or rolling the paint.

Source: *Adapted from "**Safe lead paint removal,**" Toronto Department of Public Health, no date; and "**Getting the lead out,**" Environmental Action, March/April, 1986, p. 21-22.*

Over the years, interior paint flakes and chips may have become scattered and entered the house dust.

Exterior paint chips can get into soil in the yard. Once in the soil, they can expose children playing there, they may be tracked into the house on shoes or by pets, or they may enter vegetables grown in the garden. Some older homes, especially those that have been frequently painted, may have a high level of lead in the soil simply from years of old paint gradually washing into the surrounding environment. Old roofing shingles also may have considerable amounts of lead which settled out on their surfaces over the years. Use care in removing shingles and dispose of them carefully in containers.

Although the most serious lead exposures during renovation activities usually occur when removing older materials, some other new work, such as soldering new wiring or plumbing, also can increase lead exposure. Since almost all paint contains some lead, do not spray paint unless you wear an appropriate mask (see Figure 17). See the *Renovators' Guidelines* for ways to keep your lead exposure low and to keep as much lead away from the rest of your family as possible. Be aware of the symptoms associated with low level lead contamination, such as stomach pains, anemia, constipation, general tiredness, and poor colour. If you are concerned about your exposure, see your family doctor and ask for a blood-lead test.

In addition to lead exposure during renovations, some hobby activities may increase your exposure. Lead is used in a number of hobbies, including some pottery glazing, stained glass work, soldered sculpture work, any other soldering activity, preparation of fishing sinkers or bullets, working with small lead weights in interior decorating, and using some types and colours of art supplies, graphic materials, inks, and dyes. Follow any of the "Renovators' Guidelines" (see Table 3.6) that are appropriate for your hobby.

Occupational Exposures

Much of the information presented earlier for renovators is also applicable to many kinds of occupational exposures. Lead exposures for people directly employed in the lead industry usually are handled by provincial regulations in Canada. For workers in other industries where lead is used (see Appendix G), it is important to have clean hands and a clean location for eating and smoking. These workers should also be very careful to not bring home particles of lead in their clothing, shoes, or hair, especially if there are young children at home. It takes only one teaspoon of lead dust to contaminate an area the size of a football field.

Painters may be in situations where they could receive large lead exposures, especially when they are renovating old surfaces. The dual-filter mask shown in Figure 17 should be worn when removing old paint or applying new paint. Since all paint contains some lead, avoid spraying if at all possible. Even rolling lead paint releases spatters and fumes near the worker's face.

In a situation where a lead-painted surface is being sandblasted or cut by a welding torch, extremely high levels of lead can be released and a high quality respirator must be worn to avoid lead poisoning. Check with your local labour department to determine the appropriate kind of respirator for the job you are doing. There may be specific governmental regulations that you will need to follow to protect yourself.

If you need to use any kind of personal equipment to protect yourself from lead exposure, make sure that you use it properly and carry out any cleaning or maintenance activities carefully. Pay attention to any ways you might be bringing lead dust home and make changes so that this situation doesn't occur.

Living Near a Lead Industry — Special Precautions

"Hey, honey, you let this lettuce go bad in the fridge. Why didn't you use it? You know I like fresh garden lettuce," said Nick.

Uh, oh, Natalie thought, I guess it's time to tell him instead of just throwing it in the garbage. "Nick, I know you like to grow our own food and it really does taste better, but I don't think we should eat any of it. I've been meaning to tell you. We've got too much lead in our garden to eat anything from there."

"What do you mean, we can't eat stuff from our garden? We ate it last year. My family has always had a garden. We always eat what we grow. You're crazy. We can't throw away good food. It's like throwing money away. You know the garden saves us money."

"I know, Nick. But after Sasha had his high blood test, I talked to the health department and they told me to get our soil tested. Well, the results came back a couple of weeks ago and we have too much lead in our soil to eat anything that grows there except maybe tomatoes and beans and a few other things. Even then we should wash it a lot. I think maybe we just shouldn't have a garden. I don't know what to do. It makes me really mad that for years we've been eating things that were bad for us without even knowing it. Its all because of that old smelter. There must be laws against them polluting in this way. I think we ought to get together with the neighbourhood and do something about it."

Living near a lead industry can increase lead exposures and does require some changes in your lifestyle unless that industry has complete control over all of its emissions. Even if there is good control now, most industries did not have such good control in the past, so soil and dust-lead levels in the area are probably higher than in other areas.

Figure 18: In some urban areas, industries and residences may be located quite close to one another.

The first thing to consider if you live near an industrial area is how much potential exposure there is for you and your family. It would be wise to find out your soil lead level. Depending on the level, you may decide not to have a garden or to grow only certain crops (see Figure 16). Since the dust-lead level inside your house will probably be high if your soil-lead level is high, you may need to do the decontamination cleaning described earlier. If the industry near you does not have complete control over current lead emissions, you will need to wet dust and wet mop frequently to keep lead levels under control inside the house. All the measures discussed above should be followed very carefully.

Consider getting involved in community action. You could join, or start, a neighbourhood residents' association. Educate yourself on lead health problems, local lead levels, and what the local industry is doing to reduce the problem. Make your views known to the local health and environmental departments. Get their assistance in helping you solve your problems. Tell newcomers to the neighbourhood what they should know about the local situation and how to protect themselves from lead exposures. One neighbourhood organization in Toronto (South Riverdale) was successful in getting a very large reduction in emissions from a secondary lead smelter in their neighbourhood. Their campaign included involvement with health and environmental agencies, the media, public education, and participation in a liaison committee with the industry and the government.

Figure 19: Frequent cleaning of window sills is a good "dustbusting" technique.

Conclusion

Lead is *not* something you have to live with. There are many things you can do to minimize your exposure. The more you know about lead's effects and where you are exposed, the more you can do to reduce your risks and change the situation. Choose some of the lifestyle suggestions that are right for you. Develop a personal or family plan of action. Begin to make some changes and don't stop until you have reached your lead control goal.

PART 2 THE WHOLE PICTURE

Chapter 4

From the Pyramids to the Poles

People began using lead over 6000 years ago in ancient Egypt. This use has increased steadily until the present time, with the exception of a fairly large drop in production after the fall of the Roman Empire. Today, lead is one of the most widely used metals. As a result, lead can be found throughout the environment at higher than natural background levels. In recent years, scientists confirmed this global spread of lead contamination when they discovered increased lead levels even in the polar icecaps.

Lead Characteristics

People have long known that lead has a poisonous nature. In spite of this fact, numerous applications have developed. One of the reasons lead is used extensively in common products is its availability. There are relatively large amounts of lead in the earth's crust, and because it has a low melting point, it is fairly easy to separate from the natural ores where it is concentrated.

The low melting point is just one of several valuable physical and chemical characteristics of lead and lead compounds that have resulted in their use in a large number of industrial and other applications (see Figure 20). For example, a low melting point makes lead useful in solder and casting metals. Its high density makes it useful for x-ray shielding, weights, and sound insulation. Resistance to corrosion allows lead to be used as a liner for chemical storage tanks and in electrical storage batteries, pipes, electric cable sheathing, paints, and coatings. In its tetraethyl form, lead added to gasoline improves performance by eliminating engine "knocking."

Lead is used also for packaging because of its malleability (softness and flexibility). An example is the lead foil seal on some wine bottles. This same characteristic allows lead to be easily molded into ammunition. Its low

Figure 20

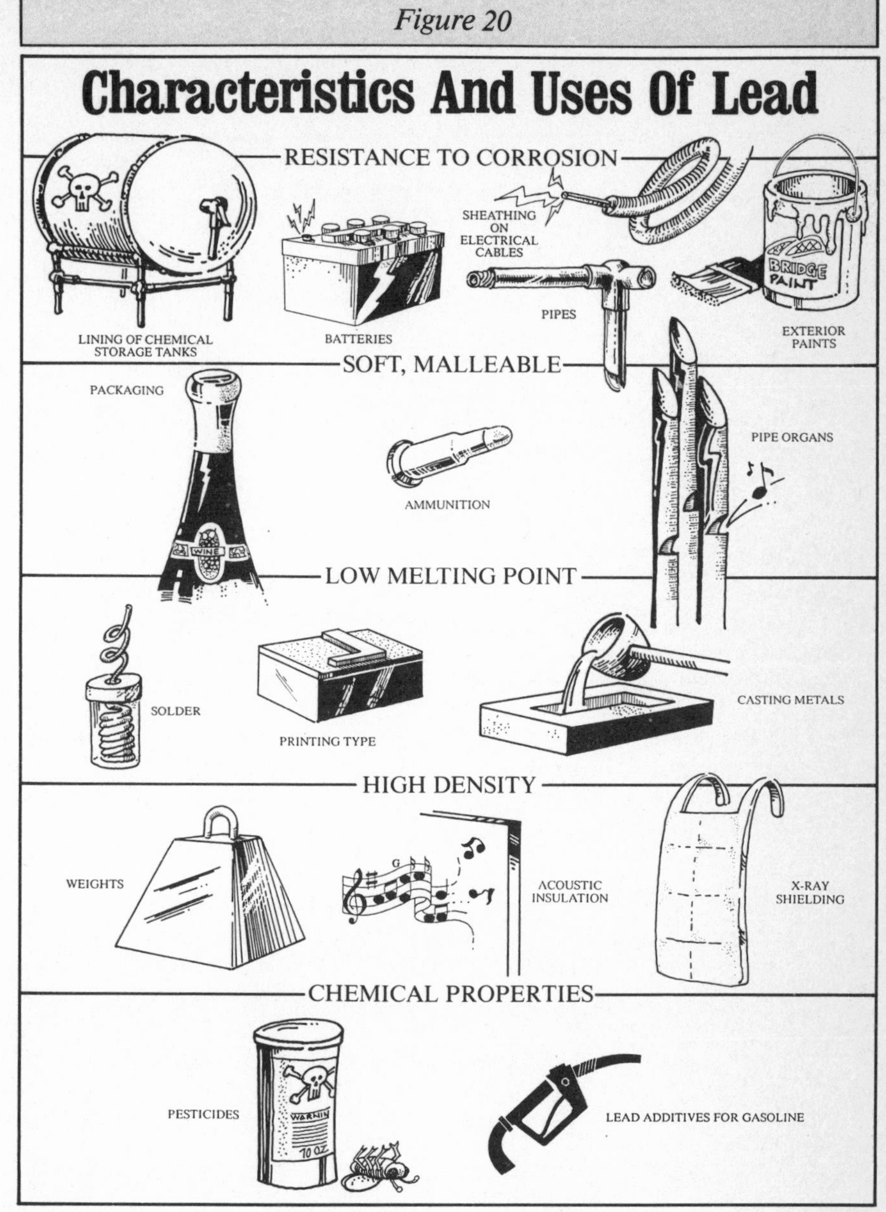

Some of the most interesting physical and chemical properties of lead and examples of the way people have used these characteristics in a variety of products.

Source: *B. Wallace and K. Cooper, Lead, People, and the Environment, Niagara Neighbourhood Association, Toronto, Ontario, 1985.*

melting point and flexibility during and after cooling have been valued for centuries by craftsmen who blend it with other metals to make pipe organs for churches. Because of the potential for harm to humans, lead is used to a much lesser extent today than formerly for pottery glazing, pesticide preparations, and dyes in coloured printed matter. For example, lead-containing pesticides were banned in Canada in the late 1970s. One of the most familiar, but incorrect uses of the word lead is in pencil "lead," which is not lead at all but graphite.

Historical Uses of Lead

People have been using lead for many of these same purposes since the time of the ancient Egyptian civilizations. The Egyptians used lead to obtain bright whites and yellows on glazed pottery and for solder, ornamental objects, sinkers for fishing nets, and plummets for measuring water depth. Other ancient Mediterranean cultures used lead for caulking, sling shot, and as writing tools, both for pencils and easily-scratched slates. In the east, ancient lead uses in China included coins and trademarks and in parts of India, trinkets and weights for looms.

It was the Romans, however, who increased the use of lead almost to the level (on a per capita basis) that is current in industrial nations today. It has been estimated that at the height of the Roman Empire about four kilograms of lead were used every year for each person.[1] (Canadian lead use in the 1980s is about five kilograms per person per year.[2] It was slightly higher in the 1970s but has declined because of environmental concerns and cheaper technological alternatives.)

A large proportion of the lead mined during Roman times was used in the building of enormous aqueducts throughout the Empire as well as for associated water pipes. Food for the upper classes was cooked and eaten almost exclusively in leaded cookware and dishes were sweetened with a lead-laced concoction called sapa.

The use of lead declined quite dramatically after the fall of Rome and began to increase again several hundred years later with the development of lead-silver mines in Germany. For most of its history, lead was not mined for its own value but because ores containing lead often also contained precious metals. Lead became a useful by-product of other kinds of mining, notably silver mining. The flourishing production of silver coins during Greek and Roman times generated a steady flow of lead. The history of lead production (see Figure 21) has, in fact, been estimated indirectly from records of accumulated silver stocks used to produce coins. With the onset of the Industrial Revolution, lead production increased considerably and lead began to be mined for its own sake.

Figure 21

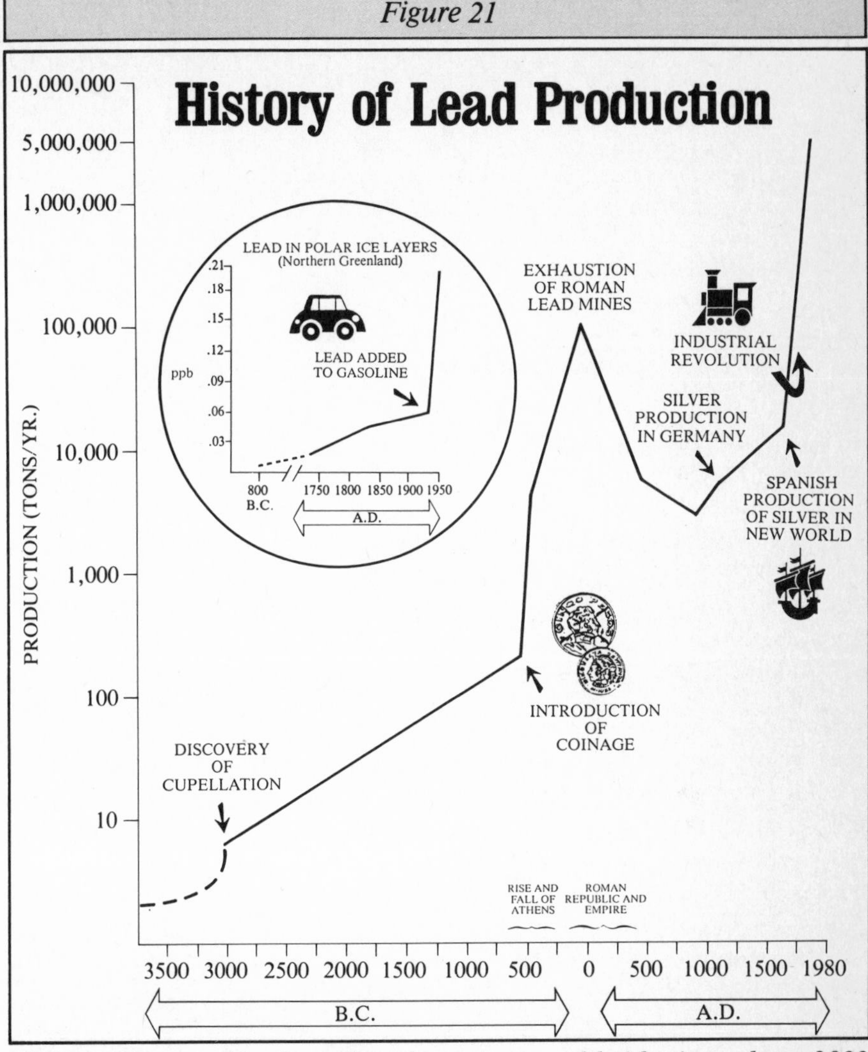

Estimated tons of lead produced per year worldwide since about 3000 B.C. The present level is about 5 million tons per year. Note that the scale for tons per year is logarithmic (each unit of increase is ten times larger than the preceding unit of increase). The insert shows measurements of lead in parts per billion in polar ice layers since about 1750.

Note: *One ton equals about 0.98 metric tonnes.*

Sources: *Adapted from D.M. Settle and C.C. Patterson,* ***Science,*** *207: 1167-1176, 1980; M. Murozumi, T.J. Chow, and C.C. Patterson,* ***Geochim. Cosmochim. Acta,*** *33: 1969, 1247-1294.*

Lead in the Polar Icecaps

In the last 200 years, lead levels have steadily increased in the remote icecap of Northern Greenland. Scientists have taken core samples of ice and snow to measure the amount of lead that was in the snow when it fell. The insert graph in Figure 21 shows a sharp increase in icecap lead levels at the beginning of the Industrial Revolution. This increase reflects a number of factors including greater overall amounts of lead use and taller smoke stacks resulting in more widespread dispersal of lead particles.

The dramatic rise in lead levels in the Greenland icecap that began about 1940 is almost entirely the result of the use of leaded gasoline in motor vehicles. This conclusion is based on the fact that the most significant increase in lead use at the time was the combination of a sharp rise in the number of vehicles and an industry-wide adoption of leaded gasoline. Lead in gasoline exhausts contaminated the snow falling on the polar icecap because the lead particles emitted in vehicle exhausts are very small and a small proportion of these particles will travel airborne for long distances. Since vehicles are mobile, their exhausts permit more widespread dispersal of lead particles than do stationary industrial locations.

The icecap in the Antarctic also has elevated lead levels in the upper, more recent, ice layers. However, the Antarctic lead levels are about ten times lower than those in the Greenland icecap and did not start increasing above natural background levels until after the 1940s.[3] This delayed increase in the Antarctic is a reflection of a number of factors including fewer cars in the southern hemisphere (it has been estimated that about 90% of atmospheric lead emissions occur in the northern hemisphere),[4] longer distances for lead particles to travel, and the long time periods involved for mixing and transfer of air masses between the northern and southern hemispheres.

In just the past 45 years or so, environmental lead levels have increased so much that they are now hundreds to thousands of times higher than natural background levels. We have used lead for many of the same reasons for thousands of years and will likely continue to do so. From car batteries to stained glass, lead plays an important role in our modern lifestyle. However, with the introduction of lead to gasoline, lead emissions to the environment have accelerated dramatically, spread by the passenger car. The use of leaded gasoline is in fact the primary cause of lead contamination on a global scale.

Chapter 5

Lead is Everywhere

It is amazing to realize that absolutely everything you can see, touch, or eat has some tiny amount of lead on it. Because of our long history of lead use, and our vastly increased rate of use in the past 50 years, we have caused global lead contamination. Although a very small amount of lead enters the surface environment naturally from deposits in the earth's crust, human lead use is responsible for over 99% of current environmental lead levels.[1]

As one of the elements out of which the earth is made, lead cannot be broken down or destroyed. It is persistent. Once it is released from the ore where it is concentrated, it will move through the environment along certain pathways depending upon its use and disposal. It can then accumulate in the environment in more easily available forms than when it was bound up in ore deposits.

Sources of Lead Entering the Environment

Although there are many different sources of environmental lead contamination, they can be divided into four categories (see Table 5.1). These categories are mobile (moving) sources, or point (non-moving or stationary) sources, household sources and natural sources. In terms of the amount of lead emitted, the two largest categories, by a considerable margin, are mobile and point sources. Lead emissions are described as the amount emitted to the atmosphere since the bulk of emissions are first released there. However, lead in the atmosphere is not a particularly important direct exposure source for people since most of the lead falls out of the air fairly rapidly.

Mobile Sources

Mobile sources include all vehicles and boats that burn leaded gasoline. These sources cause lead to be spread out over a wide area thus contributing to

Table 5.1

Lead Sources

Mobile Sources
automobiles; trucks; lawn mowers; power boats; airplanes

Point Sources Group 1: Industries Directly Involved in Lead Use
mining, milling, and smelting of lead ores; secondary lead industry (lead recovery from used lead products such as batteries, cables, etc.)*; battery manufacturing*; gasoline additive manufacturing*; manufacture of miscellaneous lead-bearing products (cables, rolled and extruded lead, alloys, etc.)*; pigments and chemicals manufacturing.*

Point Sources Group 2: Industries Not Directly Involved in Lead Use That Release Lead
copper and nickel production; zinc production; waste oil burning*; garbage and sewage burning*; iron and steel production*; coal burning*; heavy fuel oil use*; production of ferroalloys*; wood burning*; cement manufacturing*; autobody shops.*

Household Sources
burning of coloured printed matter, lead painted wood or waste oil; use of lead solder (plumbing or stained glass work); some pottery glazing; fishing sinkers or bullets; lead foil packaging; paint chips; cigarette smoke and dust; toys and figures containing lead; folk remedies; cosmetics (especially Oriental cosmetics, e.g., Surma, a black eyeliner); jewelry (painted with lead to imitate pearl).

Natural Sources
erosion of wind-blown dusts; forest fires; volcanic activity; sea spray.

*Indicates a lead source that falls into the "All Other" category in Figure 22.

Sources: *Adapted from Centers for Disease Control, "Preventing Lead Poisoning in Young Children." United States Department of Health and Human Services, Public Health Service. Atlanta, Georgia, 1985; J. O. Nriagu, "Lead contamination of the Canadian environment," unpublished report to the Royal Society of Canada's Comission on Lead on the Environment, 1985; and D. M. Settle and C. C. Patterson,* **Science,** 1980, 207, pp. 1167-1176.

extensive, low-level contamination. Nationwide in Canada, it is estimated that 63% of atmospheric lead emissions come from automobiles.[2] In cities, however, this contribution is usually closer to 80% or more.[3] In addition, for areas near mining operations, the overall Canadian estimate of 63% for automobile emissions is probably too high since these areas experience considerable atmospheric lead from non-automobile sources.

Figure 22

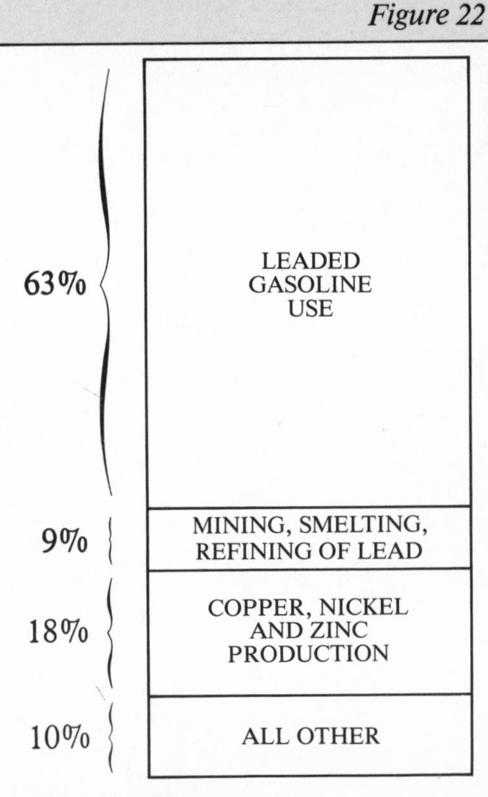

Sources Of Environmental Lead Contamination

Approximate percentages that different lead sources contribute to current environmental lead contamination in Canada. The wide variety of sources that are included in the "All Other" category are noted in the "Lead Sources" table.
Source: *Adapted from Royal Society of Canada, Commission on Lead in the Environment, "Lead in Gasoline: A Review of the Canadian Policy Issue," 1985, p. 6.*

The 60% to 80% estimate for the contribution of leaded gasoline to atmospheric lead emissions is astonishingly large in light of the fact that the production of lead additives for gasoline constitutes only about 6% of all lead uses.[4] The reason for the very large contribution of gasoline lead to the total picture of environmental lead emissions is that most of the lead in gasoline goes out the tailpipe (after it boosts the octane level of the gasoline), while most other lead uses are less dispersive.

Point Sources

In contrast to the widespread dispersal of lead by mobile sources, emissions from industrial point sources are usually fairly localized, which sometimes results in very high levels of lead contamination in the immediately surrounding area. Lead particles emitted from point sources also tend to be larger and heavier than those in automobile exhausts. The combination of larger particles that fall out of the air fairly quickly and the large amount of lead emitted from a single source can result in localized, high-level contamination. Evidence for point source contamination can be seen when there are elevated levels of lead in air, dust, soil, vegetation, and people's blood around a lead industry. These levels tend to decrease with distance away from the source.

There are a number of different kinds of industries that can be classified as point sources of lead emissions. Table 5.1 lists these in two groups. The first group of industries includes those directly involved in the use of lead while the second covers industries or activities that are not actually lead-related but that indirectly release lead in the course of other processes. It is usually only the industries in the first group that emit enough lead to cause excessive contamination in the surrounding area. Most of these industries have brought their emissions under control in recent years. However, even in the case of a currently well-controlled industry, past emissions probably were not as well-controlled with the result that there may be significant amounts of lead present in soil, street dust, or house dust in the area. This historical build-up of lead in soil and dust may contribute to elevated blood-lead levels today.

The control of emissions from lead industries located in urban areas is of particular concern since people who live or work near an urban lead industry are already exposed to high lead levels as a result of the urban concentrations of cars. In addition to whatever current emissions there may be from the lead industry, they face exposure from the historical build-up of lead in soil and dust. Because of this triple exposure possibility, it becomes very important to have stringent controls on current urban-industrial lead emissions.

The Group 2 list of point sources is somewhat surprising in that it contains industries which most people would not associate with lead emissions. For example, autobody shops may be a source of lead emissions since the exterior paint on cars often contains lead and this paint may be sanded or chipped off when body work is done (see Figure 23). These shops represent a largely unaddressed occupational hazard as well as a source of lead emissions to the local environment.

It is important to remember that when emissions from all sources are considered, the total of all point sources contributes far less to environmental lead contamination than leaded gasoline emissions. However, in a particular geographic location, a point source may be a significant contributor.

Point sources release lead into sewers and bodies of water as well as into the atmosphere. Liquid waste containing lead is required to be treated in some manner to keep its lead level below a government-specified maximum. However, some lead release is allowed. Industrial sources usually release lead in a form that is largely insoluble in water. As a result, lead particles will bind to other particles and end up in bottom sediments or in sewage sludge. The lead in bottom sediments is potentially available to small organisms that obtain their food supply from these sediments. The lead taken in by these small organisms can be passed up the food chain when these organisms are eaten by larger organisms. Lead in sewage sludge may also be released to the environment if the sludge is burned in a sewage treatment plant incinerator[5] or applied to agricultural lands as fertilizer.

Figure 23: Sanding old paint from an automobile can be a significant lead exposure source that many people are unaware of. A dual-filter mask, such as the one show in Figure 17 should be worn.

Although most industrial lead emissions are insoluble in water, this lead can become water soluble if the water is sufficiently acidic. In addition, about half the lead in gasoline lead emissions is water soluble. Solubility in water allows lead to move through the environment as effectively as water moves.

When the water in lakes or streams becomes more acidic, for example as a result of acid rain, the amount of soluble lead in the water also increases. As a result, lead from either mobile or point sources that is bound up in bottom sediments can become soluble and be released into the water.

All of this newly-soluble lead can contribute to higher lead levels in various organisms or in drinking water supplies.

Household Sources

Many of the potential sources of lead in the household are listed in Table 5.1. Paint chips represent the most common household source of lead. Ingestion of paint chips by children has caused many serious cases of lead poisoning in the past and still represents a danger. Older paints contained up to 50% (500,000 ppm) lead. With recognition of the problem and regulation of the lead content of paint in recent years, the number of cases of paint-related lead poisonings has been greatly reduced. This reduction has been particularly strong in the United States where such poisonings were a much greater problem than in Canada. Even though the current allowable level of lead in paint (5000 ppm in Canada and 600 ppm in the U.S.) is much less than in older paints, it is still high enough, especially in Canada, to cause problems if it is ingested by a child. Any paint chips should be considered as a potential source of lead and properly handled and disposed of as described in Chapter 3.

Cigarette smoking is a lead exposure-increasing activity in the household, because the lead in the smoke can contaminate house dust. Only about six percent of the lead in cigarettes is inhaled by the smoker with much of the remainder becoming incorporated into house dust.[6]

In addition to the sources shown in Table 5.1, plumbing may be a source of lead in two ways: lead-soldered joints or lead pipes. Much of the plumbing in Canadian homes consists of copper pipes with lead-soldered joints. However, many of the larger and better-built homes constructed before about 1920 contained lead plumbing pipes, which may still be in use. Lead pipes result in much higher lead levels than do copper pipes with lead-soldered joints.

In general, the amount of lead in the tap water is quite small, but can be somewhat higher at some times than at others. Testing of drinking water has shown that, especially in homes with lead pipes, the highest lead levels are found when the taps are first turned on in the morning after water has been standing in the pipes overnight. The lead content drops sharply after the water is run for a few minutes. This overnight increase in lead probably results from unmoving water sitting in contact with the lead or lead solder for several hours permitting tiny amounts of lead to move into the water. In situations where the water supply is soft and acidic, more lead will move into it. To flush out the first-draw water with higher lead levels, let the tap run for a couple of minutes before using it.

Natural Sources

The final category of environmental contamination is lead from natural sources. All lead that everyone is exposed to originally came from the crust of the earth. Some small amount of natural lead still comes directly from this source by way of erosion or wind-blown dusts, forest fires, sea spray and volcanic activity. These sources account for less than 1% of current environmental lead levels. If people had never begun mining and using lead, environmental lead levels would be very low, at what is now called the natural "background" level of lead. This natural background level for lead in the atmosphere has been estimated to be about 0.000024 micrograms of lead per cubic metre of air ($\mu g/m^3$).[7] Current air-lead levels in Canadian cities range from 0.2 to 0.7 $\mu g/m^3$,[8] which is thousands of times higher than the natural background level.

The Importance of Very Small Amounts

Substances measured in micrograms may seem to be too small to warrant serious consideration. However, this is not the case. For toxic substances like lead, it is necessary to get some perspective on these small amounts and realize their importance in potentially disrupting normal health.

The amounts of lead constantly flowing into the environment from the four source categories described above are so small that for all intents and purposes they can be considered as invisible. Although the word lead is associated with images of something substantial and heavy, or of a dull grey colour, the amount of lead that is a human health hazard may be very tiny, or essentially "invisible." Measurement of these small quantities is usually in micrograms (a millionth of a gram) or in parts per million. See Measurement Units, Appendix F, for equivalent measures.

Some examples of the importance of very small amounts of lead include:

- serious lead poisoning if 250 micrograms are absorbed daily;
- for a young child, the maximum daily absorption should not exceed 100 micrograms;
- licking a finger that has city dust on it can add several hundred micrograms to a child's intake; and
- one serving of beans from a lead-soldered can has about 35 micrograms of lead.

Since you cannot actually see this "invisible" lead, you need to be aware of ways in which you can be exposed to it. In your daily routine and during special circumstances (such as renovations), the information in this book can be used to identify potential sources and potential exposure pathways in your life. Note that although the majority of environmental lead contamination first occurs as emissions to the atmosphere, most lead falls out of the air quite rapidly and your exposure occurs as a result of where this lead eventually ends up.

Sources of Lead for People

In Chapter 3 we saw that the majority of human lead exposure occurs through food and dust. The fact that these sources are the two largest is because of the pathways lead follows in the environment. There is a continuous fallout of lead particles from the atmosphere. Other lead enters the environment from numerous small sources including flaking paint and bits of solder. The particles of lead tend to bind to other particles and accumulate in soil, dust, and dirt. Food surfaces become contaminated by lead fallout during growing, harvesting, transporting, processing, and being displayed at the grocery store. Some food crops also take up lead from the soil in which they are grown. Animals can take in lead when they eat and drink and pass it on in their milk or meat.

Although environmental lead contamination is global in extent it is obviously not distributed equally. The amount of lead you are exposed to depends upon where you live. Everyone's lead exposure from food is about the same no matter where they live. (However, people who eat much food from lead-soldered cans will have higher food lead exposure levels.) Dust-lead levels, however, increase as you move towards urban areas and higher concentrations of automobiles. Near urban lead industries, dust-lead levels can be quite high due to either poorly controlled emissions or high soil lead levels, or both. Blood-lead levels in people have been shown experimentally to increase as you move from remote to rural to suburban to urban to urban lead industry areas. A similar increase in lead levels occurs in air, soil, dust, and vegetation along this gradient.

Even if all lead were taken out of gasoline immediately and all industrial emissions were completely controlled, lead's persistence means that current levels of environmental contamination will stay with us for many years to come. It seems prudent, therefore, for us to realize that 50 years of widespread leaded gasoline use, along with poorly-controlled industrial emissions, is a very large environmental mistake that we would be wise to end as soon as possible.

Chapter 6

Damaged Health

A Poison

Lead is an element, one of the group called heavy metals. All heavy metals are poisonous at certain concentrations. They can be absorbed into the body and, in large enough amounts, can cause serious health problems, including death. Even very small amounts of substances like lead can cause health problems for humans.

Since lead must be absorbed into the blood before it can cause toxic (poisonous) effects, a person's risk of developing lead-caused health problems is gauged by the amount of lead in their blood. This measure is usually described as so many micrograms of lead for each decilitre of blood (μg/dl).

Once lead gets into the blood, it can move around to cause health problems in many different organs or systems. In children, some of the most important low-level exposure effects of lead are seen in the brain where lead interferes with normal development and activity. In adults (see below), low to moderate exposure levels are more likely to cause problems in the digestive tract, kidneys, reproductive system, or in blood pressure increases.

At very high levels in both children and adults, lead poisoning, including severe brain damage or death can occur. Although lead poisoning is now very rare, it can occur in such extensive lead exposures as sandblasting lead paint off a bridge without protective gear, or eating a large amount of old, leaded paint chips. Smaller degrees of exposure with less severe toxic effects are widespread in the general population.

Some of these effects are so subtle that they may go unrecognized, even though they may be lowering the quality of health or preventing the development of full intellectual potential. There are no unique symptoms associated with low-level health problems from lead. The symptoms that do occur are often mistaken for other illnesses, such as flu or stomach problems. Some subtle effects, particularly in children, may not be easily noticeable until they reach school age and encounter more demands on their ability to learn or concentrate.

The effects of lead can be persistent. That is, once lead has damaged a cell, or a system, that damage may last for years or may even be permanent. For example, subtle changes in electrical activity in the brain have been shown to last for years.[1] The fact that lead can cause these relatively permanent and irreversible effects is an important reason to take action wherever possible to avoid their development in the first place.

A few years ago, most health professionals thought that few were at risk of lead health problems outside of children living in deteriorated housing in inner cities. Based on the results of large-scale studies of blood-lead levels in children in Canada and in the general population in the United States, this idea now has been shown to be incorrect. Apparently (because of the many possible sources of exposure to lead) no matter where you live, you may be at risk of lead health problems.

A Threshold for Lead Effects

A threshold level is defined as the amount (in this case, of lead in the blood) below which there are no health effects and above which there are health effects. For lead, there are two different kinds of threshold levels. The first level is the threshold above which a doctor can observe clear signs and symptoms of lead problems. The second level is the threshold above which changes occur inside cells or systems in the body that may not be evident unless special, sensitive tests are made. The first type of threshold is called "elevated" blood lead and the second, "biochemical changes."

The actual level of lead in the blood that usually is associated with each of these two thresholds is discussed in the "Risks and Levels" section. The most important fact to remember about thresholds is that they are not a "real" point. The threshold simply represents a place along a continuum of changes where it is obvious that a change has occurred.

Consider, for example, an imaginary substance with a threshold of 100. There is not really a great deal of difference between 99 and 100 or 101 and 100. That is, the processes that are operating in the 99 situation are much more similar to 100 than they are to 85 or 90. Although persons with a value of 99 may breathe a sigh of relief that they are not at 100, intuitively they feel a little more anxious about their condition than those who have levels of only 60 or 70.

The somewhat artificial nature of the threshold point is taken into account, in part, by considering "individual differences." A threshold is a kind of average value that may or may not apply to a particular individual. Lead, like many other toxic substances, does not act identically in every person. Some people are more sensitive than others to its effects. For this reason, someone with a blood-lead level below the threshold may experience the same problems as someone with a blood-lead level above the threshold. And conversely, someone with a blood-lead level above the threshold may not experience the predicted, above-threshold effects. An attempt has been made to deal with these individual differences by categorizing lead health effects on the basis of the *lowest* blood-lead level that is known to cause a certain effect.

The group of people most sensitive to lead health effects is children under about six years of age with unborn children and two to three year olds the most sensitive subgroups because of their stage of development and their potential for large exposures. An unborn child (fetus) is directly exposed to the lead in the mother's blood. Therefore, women of child-bearing age are considered to be a sensitive group, although the effect is on the unborn child rather than on the woman. Another sensitive group is people who have an inadequate amount of calcium or iron in their diet, with the result that they absorb a higher percentage of the lead to which they are exposed than do others.

The thresholds for various lead effects are shown in Tables 6.1 and 6.2. The values for children and adults are shown separately because of the differences between these groups in their lead sensitivity (i.e., the rapidly developing child's system can be more easily and seriously damaged).

Risks and Levels

The Centers for Disease Control (CDC) in the United States periodically examine scientific information to determine the amount of lead in the blood that should be considered as "too much." Health departments in Canada and many other parts of the world usually follow these CDC guidelines. The "elevated" blood-lead threshold level for children, in μg/dl, has been lowered over the years by CDC from a pre-1970 level of 80, to 35 in 1977, 30 in 1978, and 25 in 1985. Currently, CDC defines 25 μg/dl in children as an "elevated blood lead level, which reflects excessive absorption of lead."[2] Although there is no set "elevated" level for adults in the general population, some authorities refer to 30 μg/dl as a level that raises concerns about adult health problems. (Note, however, that lead industry workers are not removed from their jobs until their blood-lead level reaches 50 to 70 μg/dl in Canada and the United States depending on the political jurisdiction.)

The 25 μg/dl elevated level in children is somewhat arbitrary since medical authorities do not deny that lead can cause changes in normal body functioning at lower levels. However, patients below about 25 μg/dl are not advised to have their blood-lead levels lowered by medical drugs. At higher

Table 6.1

Thresholds for Health Effects in Children

Lowest Blood-lead Level for Observed Health Effect

Blood-lead Level (μg/dl)	**Health Effect** (See text for more information)
less than 5	Inhibition of Py-5-N activity
less than 10	Inhibition of ALA-D (an enzyme involved, via ALA, in the formation of heme)
12	Decreases in a Vitamin D hormone involved in calcium processes
15	Changes in electrical activity in the brain
15-30	Decrease of 1 to 3 points in I.Q.; attention deficits; behavioural changes
30	Slower "message" speed in the peripheral nervous system
30-50	Decrease of 4 points in I.Q.
40	Increases in ALA (affecting heme and brain neuro-transmitters); decrease in hemoglobin production
40-60	Diseases in peripheral nervous system; defects in cognitive activities
50-70	Decrease of 5 points in I.Q.
60	Colic and other intestinal problems (e.g., stomach ache, constipation, diarrhea)
70	Anemia
70-80	Kidney disease
80	Severe brain damage; severe mental retardation
80-100	Death

Health effects at these blood-lead levels are reliably associated with an unacceptable risk for such effects among at least some children. Due to individual differences, dietary variations, and other factors, many other children may not show these effects until a higher blood-lead level is reached.

Key to Tables 6.1 and 6.2

ALA	aminolevulinic acid
ALA-D	aminolevulinic acid dehydrase
EP	erythrocyte protoporphyrin
Py-5-N	pyrimidine-5-nucleotidase

Source: *United States Environmental Protection Agency, "Air Quality Criteria for Lead," 1984, Vol. IV.*

Table 6.2

Thresholds for Health Effects in Adults

Lowest Blood-lead Level for Observed Health Effect

Blood-lead Level (μg/dl)	**Health Effect** (See text for more information)
5	Increase in blood pressure (not yet fully confirmed)
less than 10	Inhibition of ALA-D (an enzyme involved, via ALA, in the formation of heme)
15-20	In women, increase in EP
25-30	In men, increase in EP
30	In women, increase in stillbirths and premature births
30-40	Slower "message" speed in the peripheral nervous system
40	Increases in ALA (affecting heme and brain neurotransmitters)
40-50	Kidney disease; colic and other intestinal problems; in men, changes in testicular functioning
40-60	Cognitive and other neurological deficits
50	Decrease in hemoglobin production
80	Anemia
100-120	Irreversible, severe brain damage; death

Health effects at these blood-lead levels are reliably associated with an unacceptable risk for such effects among at least some adults. Due to individual differences, dietary variations, and other factors, many other adults may not show these effects until a higher blood-lead level is reached.

Source: *United States Environmental Protection Agency, "Air Quality Criteria for Lead," 1984, Vol IV.*

blood-lead levels, doctors "wash" the lead out of the body by a chemical process called chelation (pronounced kee-lay-shun). Since this process has some degree of risk, it is not considered advisable to use it unless necessary. There are, however, other ways to lower blood-lead levels than medical procedures (that is, through lifestyle changes). The fact that the elevated level was set at 25 μg/dl, which is usually the lower limit for using chelation treatment, appears to reflect to some extent an emphasis within the medical profession on treatment rather than prevention.

The "biochemical change" threshold is less clearly definable. As mentioned earlier, lead can affect normal functioning in many cells and systems in the body. When one or another of these effects is detected, the corresponding level of lead in the blood becomes the threshold for that kind of change. Some biochemical changes are increases or decreases in certain substances in the blood; some are changes in metabolic processes; and some are changes in the speed at which messages are conducted in the nervous system.

For each change, there is a certain blood-lead level below which that change has not been seen to occur consistently (see Tables 6.1 and 6.2). The lowest blood-lead level at which any biochemical changes have been noted is about 5 μg/dl. Most scientists consider a blood-lead level of 10 to 15 μg/dl as indicating the threshold above which a significant number of biochemical changes begin to occur.

Medical authorities have not reached agreement on the effect on health (i.e., toxicity) of these biochemical changes at blood-lead levels below about 20 μg/dl. Some say that changes below this level do not indicate health effects and that the body can continue to function normally even with these changes. For one or two of these changes (e.g., the Py-5-N inhibition discussed later), this position may be valid because the body apparently has a reserve supply of the substance depleted by lead. For other below-20 μg/dl changes, many authorities maintain that the changes themselves are valid health effects. For example, Dr. Sergio Piomelli, a prominent lead expert at Columbia University, has said that, "To discount the biochemical evidence of lead toxicity as irrelevant to health is equivalent to denying the biochemical nature of most biological processes and of life itself."[3]

In addition to the "elevated" and "biochemical change" levels discussed above, the Royal Society of Canada's Commission on Lead in the Environment has recommended, and some departments of public health (e.g., Toronto) are using, a third level called the "intervention" or "alert" level. This level in 1985 was usually 20 μg/dl. Action at this level consists of a second blood test and a home visit to determine if there is an obvious source of lead exposure that could be removed. In addition, soil tests and suggestions for lifestyle changes may be made.

This kind of "intervention" level focusses more on prevention than treatment and is, therefore, helpful in controlling lead risks. It is, however, relatively high since significant biochemical changes begin to occur at 10 to 15 μg/dl. An "allowable" level of 15 μg/dl, coupled with careful lifestyle counseling on ways to reduce exposure and absorption to lead, would probably improve the quality of life and reduce risks of lead health effects for a significant proportion of the population.

There are several blood-lead levels to keep in mind when considering health problems in children. (Note that there are no comparable levels for adults except the suggestion of 30 μg/dl as a possible elevated level.) **The elevated level, requiring medical intervention, is 25 μg/dl. Where used, the allowable level, requiring blood re-testing and inspection of possible exposure sources, is 20 μg/dl. The biochemical change level, indicating the existence of a number of internal bodily changes, is 10 to 15 μg/dl.**

The Domino Effect[4]

The most important change that begins to occur at the biochemical threshold is the reduction of a vitally important substance in blood called heme (pronounced heem). One of heme's jobs is to unite with a protein called globin to form hemoglobin, which carries oxygen in the blood to all parts of the body. Without an adequate oxygen supply, cells cannot function normally. Heme is also a necessary component of other blood proteins that are involved in the normal functioning of cells in all tissues of the body. Heme is like a kingpin in the body. When the amount of heme is decreased, a chain of events begins that cascades throughout the body like a row of dominoes falling over. Many of these "domino" effects are shown in Figure 24.

Full-blown heme reduction effects can be seen at blood lead levels above 30 μg/dl. However, the beginnings of disruption of the heme system can be seen at blood-lead levels of only 10 to 15 μg/dl. And, anywhere above this point, as the level of lead in the blood rises, there are larger decreases in heme and larger disruptions in the normal functioning of the body.

Decreases in heme are usually measured by observing changes in a number of other body substances which act as heme indicators. Two indicators that show clear changes at or below 15 μg/dl are neurotransmitter chemicals in the brain and an important substance in red blood cells called erythrocyte protoporphyrin, or EP. Changes in brain neurotransmitters are thought to be associated with lead-caused decreases in intellectual ability.

A large number of undesirable shifts in health and well-being occur when the heme domino effect comes into play. For example, heme's effects on the liver mean that the body may find it harder to detoxify drugs and other foreign substances. Throughout the system, structural damage occurs inside cells resulting in impaired energy flows in the cell and increased need for oxygen. Changes in brain chemicals involved in message transmission mean that information of all types may be communicated inaccurately or with some other distortion.

In addition, heme reduction can set into motion some Vitamin D changes. At low blood-lead levels there is a disruption in the synthesis of an active Vitamin D metabolite.[5] Because of this disruption, there is a smaller amount of usable Vitamin D available for normal body functions.

The most important role of Vitamin D is to facilitate the body's use of calcium. Adequate amounts of calcium are necessary for cells, tissues, and organ systems to function normally. If there is not enough calcium, muscles (including the heart muscle), nerves, and the endocrine system will be affected. Calcium is also used in the formation and development of bone and teeth.

Low blood-lead levels also inhibit an important enzyme (Py-5-N or pyrimidine-5-nucleotidase) that is necessary for the normal functioning of red blood cells. It plays a role in their maturation, activity, and survival.

Figure 24

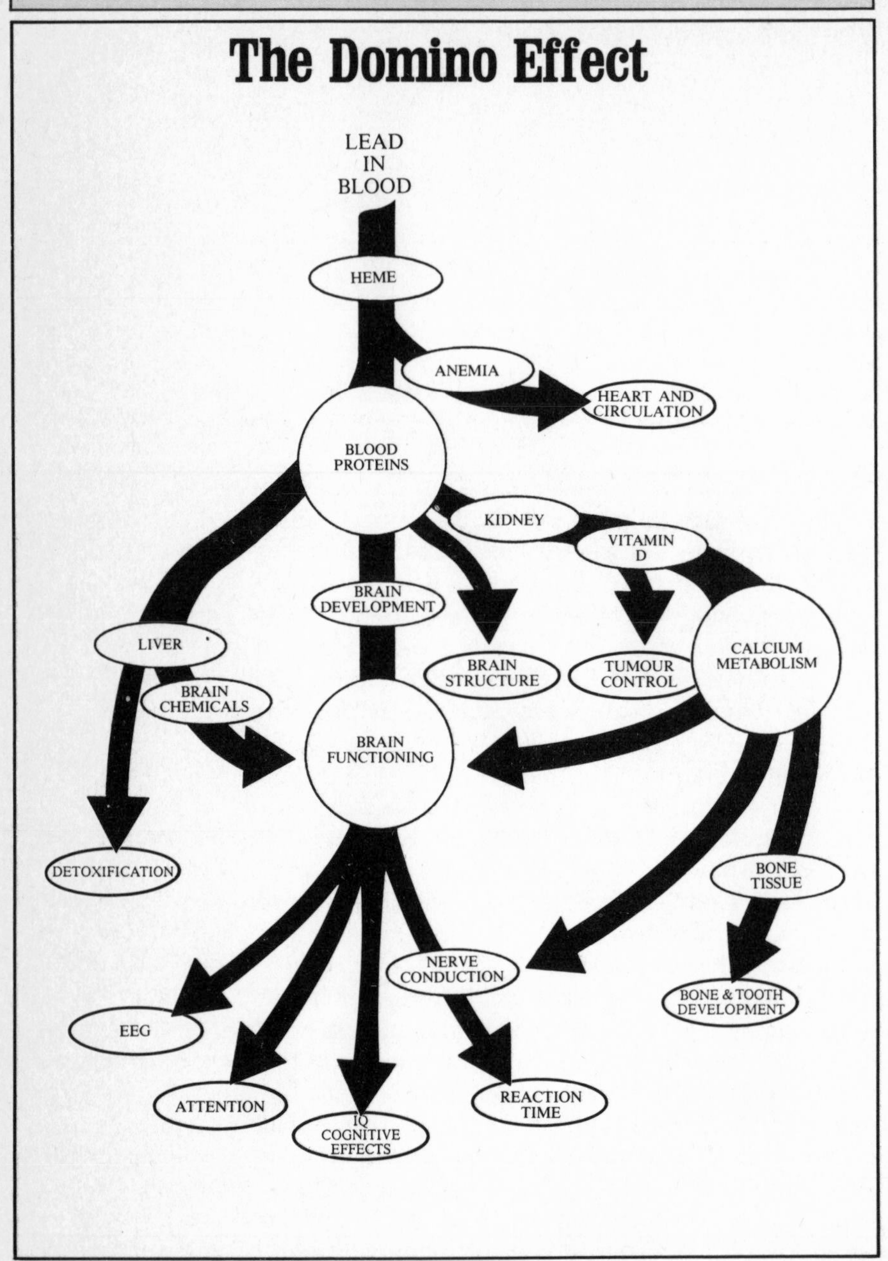

The many impacts of lead-caused reductions in heme. See text.
Source: *Adapted from U.S. Environmental Protection Agency, "Air Quality Criteria for Lead," (EPA-600/8-83-028B), 1984, p. 1-138.*

Py-5-N begins to decrease at blood-lead levels below 5 μg/dl, but, because there is a reserve supply of this substance, its disruptive effects on red blood cells are not usually seen until the blood-lead level reaches about 30 μg/dl. Py-5-N's disruptive effects cause red blood cell membranes to become more fragile and the cells to have a shorter lifespan. As lead levels increase and the amount of Py-5-N decreases, these red blood cell changes can cause decreases in hemoglobin, which may lead to anemia. The hemoglobin decrease is clearly detectable at blood-lead levels of about 40 μg/dl in children and 50 μg/dl in adults.

Children's Intelligence[6]

In children, beginning at blood-lead levels between 15 and 30 μg/dl and increasing at levels above 30 μg/dl, a number of changes in normal brain functioning have been observed. These include:

- a decrease in I.Q. from one to four or so points;
- slower reaction time;
- a decrease in the ability to focus attention; and
- inappropriate social behaviour.

The changes in brain neurotransmitters (mentioned under heme effects) cause distortions in the normal transmission of information. In addition, there are persistent abnormalities in electrical activity in the brain, including slow wave changes and reduced electrical response to auditory signals. And if lead is present early in life, there may be a delay in development of the nervous system and even destruction of developing neurons due to reductions in important brain proteins.

In the last few years, there have been a number of studies investigating the effect of low levels of lead on intellectual functioning in children. Consistently, these studies have found evidence of cognitive and behavioural deficits at lead levels that formerly were considered "safe." Because of the significance of these findings, there has been considerable scientific debate about these studies. Cognitive functioning is strongly affected by heredity and a number of other variables such as parental education, opportunities (usually described indirectly as socio-economic status), nutrition, and so on. In order to measure accurately the effects of lead, all of these other factors must be controlled experimentally.

Several recent experiments have attempted these kinds of controls. One of the most interesting of these studies was conducted by Dr. Herbert Needleman in 1979 on over 2000 children in first and second grades in Massachusetts.[7] Children at this age begin to lose their baby teeth. Since levels of lead in teeth are a good indicator of lead exposure over the preceding years, tooth lead analyses can be used to conveniently divide children into high and low lead groups. Using this approach, 58 children with high lead levels and 100 with low levels were selected for detailed evaluations, including I.Q. tests, other neuropsychological tests, and ratings of classroom behaviour. The results showed that children with higher lead levels had statistically significant lower I.Q.'s (about four points), deficits in attention and language processing, and more negative ratings on classroom behaviour.

Because of the importance of this study, it was very closely examined by a number of critics, who made suggestions for other statistical analyses that they felt would improve the validity of the original study. When these new analyses were carried out, there were no changes in the statistical results or the conclusions. The results of this study, in combination with the results of a number of other, similar studies, have led most of the scientific community recently to accept the fact that low or moderate lead levels are capable of decreasing children's I.Q.

The significance of this lead-I.Q. finding is potentially very important. Although the difference between an I.Q. of 105 and 100 may not seem very great, such differences are important especially at the upper and lower ends of the I.Q. distribution. A shift of about five I.Q. points may be enough to limit severely the number of geniuses and near-geniuses and to double the number of mentally retarded children.[8] The results of a lead-caused, across-the-board lowering of intelligence would be felt throughout society in such diverse areas as a reduction in the number of technological breakthroughs and increased needs for remedial educational facilities.

A further topic of considerable interest to present generations is the current I.Q. level. Since lead in gasoline is the largest known single determinant of blood-lead levels in the general population, and since leaded gasoline was used most extensively from the mid-1940s to about 1980, it is possible that many people born during this time span might have had slightly higher I.Q.'s if there had been less lead around. Since we have now become aware of this link between low lead levels and I.Q., strong measures for reducing the amount of lead in the general environment and the amount that individual pre-school aged children are exposed to would seem wise.

Adult Health Problems

Exposure in adulthood does not result in the kind of intellectual deficits from lead exposure that can be seen in children, except in the case of high lead levels where there is actual damage to brain tissue. Adults do, however, have problems in the peripheral nervous system (i.e., those parts of the nervous system beyond the brain and spinal cord, such as nerves in the arms and legs). As shown in Table 6.2, slower conduction (movement) of messages in the peripheral nervous system begins to be seen at about 30 μg/dl. As the blood-lead level increases above this point, this slower message transmission can result in an unsteady walk or poor coordination of voluntary muscle movements.

Pregnant women have a slightly greater risk of having a premature birth or a stillborn child if their lead level is above 30 μg/dl. Some inconclusive evidence also suggests the possibility of minor birth defects showing up at about this lead level. At about 40 μg/dl, males may develop some problems in testicular functioning, such as fewer, weaker, and more deformed sperm than normal.

As discussed under the "Domino Effect" above, erythrocyte protoporphyrin (EP) is an important substance in the blood associated with the quality of red blood cells. Women begin to show a decrease in EP at less than 20 μg/dl and men at about 25 to 30 μg/dl. At higher blood-lead levels, these EP decreases are associated with the development of anemia and they could impair heart functioning and circulation in the body.

In very recent years, a startling relationship has been found between low blood-lead levels and high blood pressure (hypertension) in adult men. For almost a century, it has been known that high blood-lead levels were associated with hypertension. It has only been within the past few years that evidence has been accumulating on the effects of low blood lead levels on blood pressure. One of the strongest pieces of evidence was found in data from a very large study in the United States.[9] This study found a direct relationship between increases in blood pressure and low blood-lead levels in men. The relationship was present even at blood-lead levels as low as 5 μg/dl. This finding is of great importance because of the essential role that hypertension plays in stroke and heart disease, two of the leading causes of death.

High Lead Levels

Above 40 μg/dl in children and 50 μg/dl in adults, blood changes become more pronounced with clearly detectable reductions in hemoglobin that, if unchecked, will lead to anemia. At about the same level, adults begin to develop problems with their digestive systems and can show signs of kidney disease.

As blood-lead levels increase, I.Q. deficits in children become larger and other cognitive problems more noticeable. In both adults and children with blood levels above about 60 μg/dl, there is the possibility of brain damage. Above about 80 μg/dl in children and 100 μg/dl in adults, brain damage becomes severe and may, in some rare cases, cause death. Usually the lowest blood-lead level at which death occurs is about 100 μg/dl in children and 120 μg/dl in adults. Signs and symptoms of high lead levels include clumsiness, poor coordination of voluntary muscle movements, weakness, changes in consciousness, abdominal pain, persistent vomiting, and constipation.

The average adult who is not exposed to lead on the job is unlikely to develop any of these major lead health effects. However, as discussed in the following chapter, there are some non-routine activities that can result in a sharp increase in lead levels and may result in serious health problems. The most likely of these non-routine, lead-increasing activities are home renovations, vehicle repainting, and burning of leaded materials. Periodically there are reports of lead poisoning occurring because of such exposures.

For children, however, the picture is different. There are a number of normal activities that could result in significant increases in blood-lead levels. The special problems associated with lead exposure in children was discussed in Chapter 3.

Chapter 7

Exposed at Home, Work, and Play

Everyone is exposed to lead from four basic sources: food, dust or dirt, inhaled air, and drinking water. Estimates of the amount of blood lead from each of these sources can be found in Chapter 9. Some people have additional exposures from their occupations, hobbies, or other lifestyle factors. In addition, living in an urban area, near a lead industry, or near a major highway increases the amount of lead in dust and air.

Children: The Special Situation

As discussed in Chapter 3, children from about one to six years have the highest exposure to lead of any age group and are the most likely part of the population to experience problems with lead (except for some adults who may experience high exposures due to their occupation or some unusual situation).

The primary reason for high lead exposures in children is their lifestyle. Children frequently play in yards or playgrounds, on streets or other outdoor areas, or indoors on the floor. These areas are all surfaces that are more likely to have heavier concentrations of lead particles than are surfaces encountered by adults in their usual daily activities.

In the normal course of the day, young children tend to get dirty. Very young children frequently suck their fingers or toys or engage in other hand-to-mouth activities. They often explore new objects by seeing how they taste. Some children develop a habit of eating non-food items. All these activities can result in significant increases in lead exposure.

Another major factor in high lead exposures for children is their height. Because children are shorter than adults, the air they breathe is closer to the floor or ground than is the air adults normally breathe. This "lower" air, particularly in traffic situations, is likely to contain more lead dust and particles than higher air.[1]

The smaller body size of a child as compared to an adult (e.g., two-year olds weigh about 10 to 13 kilograms and adult males weigh about 70 kilograms) means that the same amount of lead will be more concentrated in the child's smaller body. For example, if both the adult and child consume about 20 micrograms of lead in their diet, the equal intake would result in an unequal concentration in the body when it is expressed as lead per kilogram of body weight. In this example, the child would have 2.00 micrograms of lead per kilogram while the adult would have only 0.28 micrograms of lead per kilogram. Typical food-water consumption amounts and body weight differences result in young children taking in about three times more lead per unit of body weight than adults.[2] Because of the smaller size of children, a given amount of lead is more likely to cause problems.

A further factor contributing to higher lead exposure for some children, as compared to most adults, is their typical diet. Children who tend to eat more canned juices and processed foods will be more exposed to lead than others who do not eat as much of these items. Some other higher exposure factors for children are discussed in Chapter 3.

Living in a City

Many of the major sources of lead exposure tend to be concentrated in or near cities. The most important of these is traffic. As was discussed in Chapter 2, the exhaust of vehicles that use leaded gasoline is the largest environmental source of lead. And vehicle use is highly concentrated in and around cities. The effect of vehicular traffic concentrations on soil-lead levels can be seen clearly in Figure 25. Elevated soil-lead levels occur throughout the downtown core of a large city (Toronto) and extend like fingers along the major traffic corriders connecting the city to the surrounding area.

The soil-lead pattern shown in Figure 25 is just one measure of the higher levels of lead in a city. Airborne lead that contributes to the elevated soil levels also contributes to higher lead levels in inhaled air, in the dust in streets or homes, and in food grown in city gardens.

Industries that use lead are also often found in or near cities. For example, in Figure 25, two of the three "hotspots" are around secondary lead smelters located in Toronto. (The third "hotspot" is a now-relocated battery manufacturing facility.) These smelters primarily recycle lead from used automobile batteries. This type of industry is usually located in or near a city, because its source material (used batteries) and its market (battery manufacturers and other producers of lead-bearing products) are located primarily in cities.

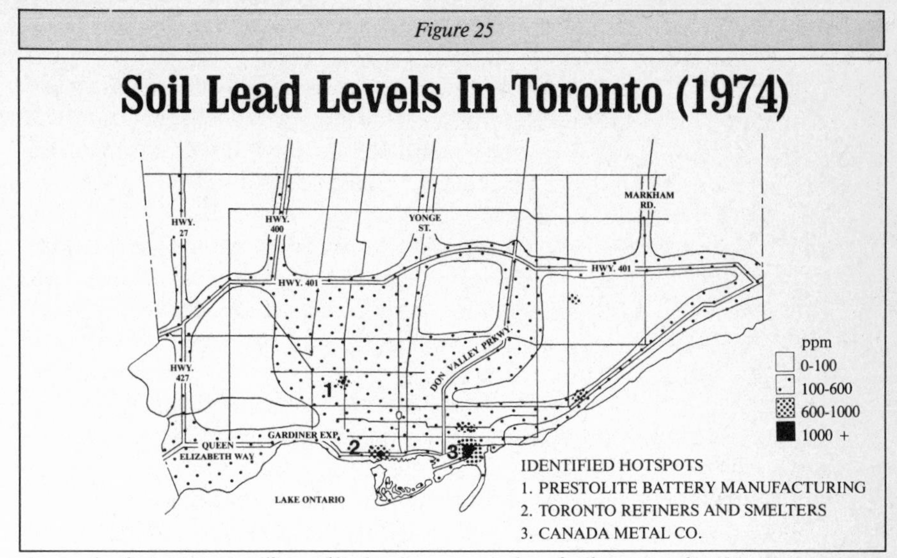

Average levels, in parts per million, of lead in the top two inches of soil as reported in 1974. Hotspot 1 is no longer in existence at this site. Hotspots 2 and 3 are lead smelters that primarily recycle used lead products, such as automobile batteries. Note the higher lead levels in the downtown core and along major traffic routes.

Source: *Adapted from K.H. Sharpe et al, "Studies of the relationship of environmental lead levels and human lead intake," Working Group on Lead, Ontario Ministry of the Environment, 1974.*

Cities also have a large number of other lead-handling establishments. Because several colours of automobile paint contain lead (predominantly whites and yellows), any business involved in repairing, sanding, or repainting automobiles, such as body shops, may be releasing lead to the environment. Other commercial sources often located in or near urban areas include businesses using solder or exterior leaded paint, iron and steel industries, and cement manufacturers.

In general, the farther away an area is from the centre of a city or from a lead-using industrial zone, the lower is the level of lead in the environment. The effect of reduced environmental levels away from city centres can be clearly seen in the comparison of blood-lead levels in children in urban, suburban, and rural areas.

Living Near a Lead Industry

People who live near a lead industry may have significantly greater exposures to lead than people living in other areas. If the industry is located in a rural area, the presence of the industry may offset or more than offset the usual rural advantage of lower lead exposures. If the industry is located in a city, it will most likely cause an increase in the already high city lead levels. The population group that experiences the greatest lead exposure is children living near an urban lead industry.

Industries contribute to lead exposures in the surrounding area in a variety of ways. Around a lead mine, there is lead dust from the mining and milling operations and from piles of unstabilized tailings (the loose material left over after the valuable ore has been removed). This dust can move into the surrounding air, water, or soil.

Around lead smelters, there are two classes of emissions: stack emissions and fugitive emissions. Stack emissions refer to anything that comes out of the smoke stack. Lead smelters are required to duct (channel) the air from their smelting operations (i.e., melting a lead mixture in a furnace) through a lead-removal system, called a baghouse, before it enters the stack. Well-designed and well-operated baghouses normally remove almost all (99.9%) of the lead in the exhaust gases. Except in unusual circumstances, stack emissions are quite small. However, this is often not the case with fugitive emissions, which can be considerable. Fugitive emissions are all emissions, other than from the stack, which arise as a result of industrial operations. In the case of a secondary lead smelter, they include dust and particles from battery crushing operations, from mixing of materials to go into the smelter, from winds blowing through the premises, and from all smelting operations that are not completely ducted through the baghouse.

Since lead particles are heavy, these industrial emissions, except for the very smallest particles in stack emissions, normally have their greatest effect on the immediately surrounding neighbourhood and to a lesser extent on a somewhat larger area. In some cases of exceptionally dirty or large industrial operations, higher than normal lead levels in soil, dust, and vegetation can be observed up to five kilometres from the site with the major increases occurring within two kilometres.[3] Recently, around smaller or relatively cleaner facilities, such as the secondary lead smelters in Toronto, slight increases in environmental lead levels have been observed at a distance of a kilometre or so, while more significant increases have occurred within the first few hundred metres.[4]

Industrial lead emissions affect neighbouring residents primarily by increasing the level of lead in street dust, house dust, leaf litter, and soil. The amount of lead in the surrounding air may also be slightly increased. People become exposed to lead in soil by eating vegetables grown there or, in the case of children, by inadvertently breathing or swallowing tiny particles of lead while playing outside. Burning leaves can be a very large exposure source since a large portion of lead fallout can end up on growing or fallen leaves.

In addition, lead in soil and street dust that is brought into the house on people's shoes or feet, on toys, or by pets can increase house dust-lead levels. One important pathway between lead in house dust and human exposure is the invisible lead contamination of dishes, cooking utensils, and food that can occur during preparation or consumption. Chapter 3 describes a number of ways to reduce exposures for people living in a city or near a lead industry. Successful pressuring actions to reduce exposures for residents living near two urban lead industries are described in the section below, *The Toronto Lead Smelter Controversy.*

Occupational Exposures

A worker employed in a lead industry (i.e., mine, smelter, processing facility, transport, or battery manufacture) expects to be exposed to higher levels of lead than the average person and is regularly monitored to determine if his or her exposure is within occupational limits set by governmental agencies. Since lead industry workers are routinely exposed to higher lead levels than the general population, they could especially benefit from adopting personal lifestyle risk reducers.

Some workers in other fields are also exposed routinely to higher than average levels of lead (see Appendix B). Some of the activities and industries which regularly use lead include welding, soldering, salvage work, structural metal painting, automobile painting, some other painting, leaded gasoline manufacture, manufacture of food cans, printing (lead dyes and inks), automobile manufacturing, plastics, rubber products, stained glass, arts and crafts, oil recycling, automobile mechanics, housing renovations and firing range facilities. In some cases, workers in these industries may experience greater lead exposures and bring home more lead dust than workers directly employed in a fully regulated and cleanly operated lead industry.

The Toronto Lead Smelter Controversy

For over 40 years, two secondary lead smelters have released lead into the environment in the South Riverdale and Niagara neighbourhoods of Toronto. Several cases of lead poisoning occurred in the 1960s and children continued to be hospitalized into the early 1970s. Until residents began pressuring the government for action in the early 1970s, very little was known about the excessive lead exposures and associated problems that were occurring in these neighbourhoods.

The government did not test for lead in either neighbourhood until 1972 when they received a complaint about noise and heavy dust fallout in a Niagara Street backyard located behind the Toronto Refiners and Smelters battery crushing operation. Testing revealed astronomically high levels of lead ranging from 22% (220,000 ppm) to 43.3% (433,000 ppm) in dust on the barbecue and picnic table in this yard.[5] (Note that ore with 5% lead is considered minable.) The battery crusher was ordered shut down temporarily by the Ministry of the Environment and dust levels were reduced to between 2 and 7% lead content within a few months — still very high but considered by the Ministry at the time to be an adequate improvement.[6] Blood-lead testing revealed that many area residents had levels over 40 μg/dl, but they were assured by the government that only a level of 80 μg/dl or higher was cause for concern.[7]

Independent testing by scientists from the University of Toronto of lead in blood, hair, soil, air, vegetation and dust in the two neighbourhoods and in control, non-smelter areas, revealed excessive, localized contamination around the Niagara Neighbourhood smelter and for the first time, revealed the even larger problem around the South Riverdale smelter. Although now acting somewhat more responsibly, for several years in the 1970s the Ontario Ministry of the Environment did not take the situation in these neighbourhoods particularly seriously. They neither monitored the situation adequately nor demanded sufficient control of the smelter's operations. Repeated requests by the residents for information were often denied or responses were minimal and without sufficient explanations.

The controversy continued for seven years. During this time in Ontario and across the country, existing lead standards were revised and new standards and guidelines were set. In Ontario several large investigations were conducted culminating in a public hearing in 1976.[8] The hearing was intended to resolve the ongoing controversy about what was a safe level of exposure and how much of a clean-up was required around the two Toronto smelters. Residents criticized the hearing as a stall tactic, refused to participate, and maintained that the information that had already been gathered was sufficient to call for an immediate clean-up.

The major recommendation of the hearing board that immediately affected these neighbourhoods was the setting of 2600 ppm as a soil removal guideline.[9] Although the companies were held financially responsible for the clean-up, they ended up paying for only a third of the costs with the balance covered by a Wintario lottery donation.[10]

Over the years since the controversy began, both companies have spent considerable amounts of money on pollution control. The Canada Metal smelter in South Riverdale has become a model corporate citizen by responsibly bringing its emissions under control. These changes at Canada Metal were largely due to the effects of sustained citizen actions, protests, and negotiations with government and the industry. Since emissions from Canada Metal have largely been brought under control, and since elevated blood-lead levels continue to occur within the 1000 ppm soil-lead isopleth (area), soil removal and replacement now appears to be the most logical and permanent solution to the problem.

The situation around Toronto Refiners and Smelters in the Niagara

Neighbourhood is somewhat different. After the improved emission control and soil clean-up achieved in the late 1970s, citizen action on the lead issue lessened considerably. After a few relatively "clean" years, lead levels in the air and dustfall around the smelter again increased to unacceptably high levels in the early to mid-1980s.[11] At this point, citizen action remobilized with emphasis on emission control, safe decommissioning in the event that this company relocates, and reduction of exposure sources including contaminated soil and lead in gasoline. By 1985, the residents' organizations in the two neighbourhoods had once again united on lead issues of common concern (soil and gasoline lead).

Although it has taken many years and often a good deal of frustration and disillusionment, local residents have shown the value of public participation in mobilizing government to take appropriate action and in modifying local industrial operations. Through determination and commitment of personal time and resources, residents in both communities have made and continue to make a considerable impact on government and local industries.

Chapter 8

Lead Inside Your Body

Absorption into the Bloodstream

Although you may be exposed to relatively large amounts of lead, not all of that lead will be absorbed into the blood. In fact, some of the lead to which you are exposed may never even make it inside your body if you carefully practice some risk reducers, such as washing your hands before eating. In addition, much of the lead that does get into your body passes through unchanged and is excreted by the kidneys and bowels. The amount of lead taken in minus the amount excreted represents the amount that is absorbed.

Except for the lead in gasoline, which can be absorbed directly from the liquid gasoline into the blood through the skin, the two intake paths are inhalation (breathing) and ingestion (swallowing). The amount of lead taken in by inhalation depends on the size of the lead particles in the air and the kind of breathing that is occurring. Smaller particles and a slower breathing pattern result in a greater intake.[1] On a body weight basis, children tend to inhale more air than adults.[2] Once a particle of lead is deposited in the lungs, it is absorbed directly into the blood. About 40% of inhaled lead in both children and adults is deposited and absorbed.[3] Almost all of the lead that is not deposited is exhaled except for a small fraction that is swallowed and becomes part of the lead taken into the stomach.

Lead in dust, dirt, food, and water is taken into the stomach and intestines (both locations are referred to here as the gastrointestinal or GI tract), where differing amounts are absorbed depending on the amount of food present in the GI tract and on the person's age. More lead is absorbed if there is little food in the GI tract.[4] A child absorbs about 50% of the lead in his food, while an adult absorbs only 10%. For dust lead, the child and adult absorption percentages are 30% and 10%, respectively.[5]

Figure 26

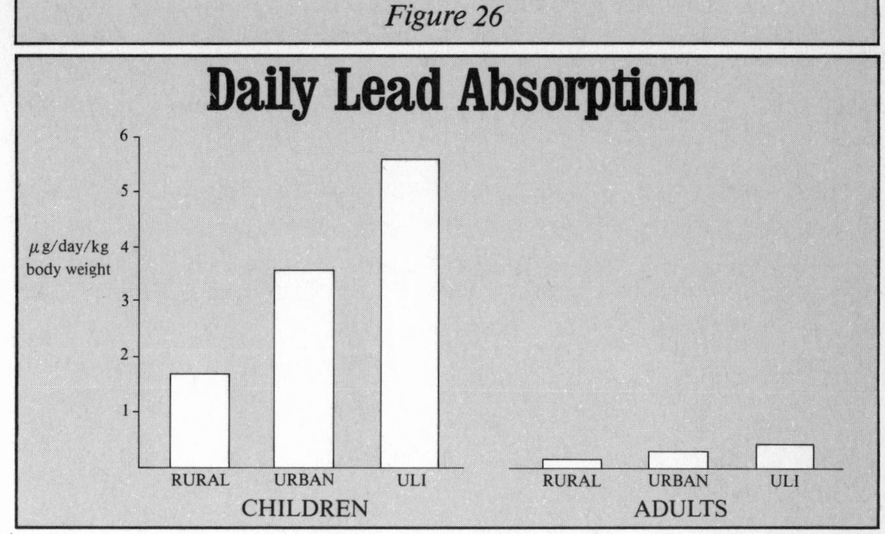

Estimated amounts of lead absorbed each day in micrograms per kilograms of body weight for one to three year old children and adults living in rural, urban, and urban lead industry areas.
ULI = urban lead industry
Body weights used: *13 kg for children and 70 kg for adults.*
Source: *Appendix B.*

Details on the estimated amounts of lead intake and lead absorption for food, dust, air, and water for children and adults living in rural, urban, and urban lead industry areas can be found in Appendix G. These data are summarized in Figure 26 "Daily Lead Absorption", which shows the total amount each group is estimated to absorb daily on a microgram per kilogram basis. As can be seen in this figure, children absorb considerably more than adults. In addition, children living in urban areas absorb more than twice as much as rural children, while children in urban lead industry areas absorb almost three and one-half times as much as rural children. Note that all these figures are estimates that can be reduced by following the risk reduction guidelines.

Measuring Lead in the Blood

There are two ways to obtain blood in order to make a direct measure of the amount of lead it contains. The first and most frequently used way, especially in children, is a simple finger prick. The second way is to take it from a vein, usually in the arm. Both methods may give inaccurate results if the person has a low iron level (is anemic). Usually the venous (from a vein) sample is more accurate than the finger prick (or capillary sample), because there is a much stronger possibility that lead on the surface of the skin will contaminate a finger prick sample. In fact, the standard procedure when a high lead level is found on a finger prick sample, is to repeat the test using a venous sample. Frequently the second, venous sample will be lower indicating that, even though the person's finger had been cleaned before the sample was taken, not all the surface lead was removed. Another possible reason for a lower venous reading is the fact that there is more hemoglobin (and, therefore, more potential for lead to be measured) in a capillary sample than in a venous sample.

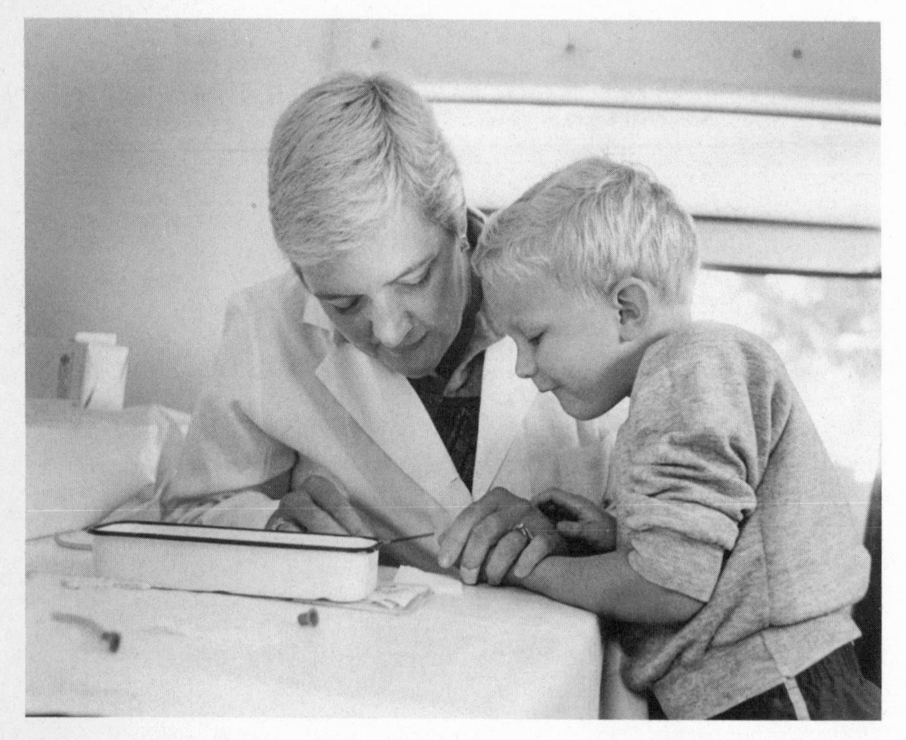

Figure 27: A child getting the simple finger-prick test that is usually used in blood lead screening programs.

Both of these methods for measuring lead in blood reflect lead exposure over the preceding one to two-month period. A third measurement technique reflects lead exposure over a longer time span, the preceding four months. This third method does not measure lead directly but measures an indicator substance in the blood called EP (erythrocyte protoporphyrin). When lead levels in the blood go up, EP levels also go up. Since the EP technique does not measure lead, there is no problem with the possibility of surface lead contaminating the sample and inaccurately elevating the measurement. An EP measurement can be taken either by a finger prick or venous sample. As with the other two measures, however, EP too is inaccurate if the person has low iron levels. In addition, an EP measurement may be inaccurate if the person has a cold, ear infection, or other minor illness. It is also insensitive at blood-lead levels below about 25 micrograms per decilitre (μg/dl).

The amount of lead in the body may also be determined by the measurement of lead in teeth, bone, or urine. Theoretically, lead could also be measured in hair, but it has not been possible to develop an accurate procedure for hair lead analysis because of the problem of lead contamination on the exterior of the hair.

Tooth lead is an excellent, but expensive, way to obtain a measure of long-term lead exposures. This method can also be used on teeth from skeletons thus providing information on lead levels in other times. Figure 2 in Chapter 1 shows the increase in tooth lead levels since prehistoric times. Measuring lead levels in shed baby teeth is sometimes used in studies relating past lead levels to school performance or I.Q.

Lead can also be measured in bone (usually available only at autopsy) and in urine. Urine lead levels are usually quite low and relatively inaccurate except when the person is undergoing medical treatment (i.e., chelation) to "wash" high levels of lead from the body.

Several different units of measure are used to describe blood-lead or EP-lead levels. The most common of these are micrograms per decilitre (μg/dl), micromoles per litre (μmol/L), and micrograms per 100 millilitres (μg/100 ml). Occasionally parts per million (ppm) or parts per billion (ppb) are used. The relationships between these various measures are shown in Appendix F. Often the results of a blood lead test are reported in μmol/L, the standard metric unit.

Where Lead Goes in the Body

When lead first enters the blood stream, it is in the plasma and then later becomes incorporated into red blood cells. At blood-lead levels found in the general population, almost all of the lead is in the red cells (90-99% compared to 1-10% in the plasma). At high blood-lead levels (e.g., 50-60 μg/dl), the amount in plasma may increase up to 65%. Only the lead in plasma is capable of moving from the blood into soft tissues (such as brain, kidney, and liver), and into bone.[6]

There is usually no problem associated with lead storage in teeth or bone, unless the amount is quite large. Lead in these locations causes problems primarily when it comes back out of bone and into blood.

The most serious problems with lead inside the body are lead's effect on heme in the blood and lead in soft tissues, such as the brain. The percentage of lead that moves into soft tissues is higher in children than adults. This percentage difference is unfortunate because it increases the likelihood of problems for the most sensitive group, the children. About 25% of new (i.e., recently absorbed) lead in children moves into more easily damaged soft tissues and about 75% goes into relatively safe bone storage. For adults, only 5% to 10% goes into soft tissue with the remainder into bone.[7] Since children's soft tissues, especially the brain, go through considerable development after birth, lead in these tissues can cause serious and sometimes irreversible effects.

The percentages in the preceding paragraph refer to what happens to new lead coming into the body. However, lead health effects are based on the total amount of lead in the blood. This total amount is a combination of new lead and "old" lead that may be coming back into the blood from storage sites in the body. Because of the role "old" lead can play in current blood-lead levels, it is important to know how lead moves around inside the body. This movement is relatively predictable depending on the rate of new lead coming into the blood. The body tends to establish an equilibrium among the amounts of lead in blood, bone, and soft tissues. When the amount of new lead increases, there will be more lead in the blood at first and then increases in soft tissue lead. If the increase is fairly large and continues, lead levels in bone will also begin increasing.

When there is a reduction in new lead, the amount of lead in the blood may not show much change for a few weeks or months until a new equilibrium is reached in all of the different lead storage locations. Gradually internal lead levels will decrease in soft tissues and blood. Normally lead that has been stored in bone is relatively permanent. The exception to this generalization occurs in people over about age 50 or so who begin to lose some of their bone mass. Lead that was stored in the bone can move back into the blood and from there go into soft tissues.

When chelation therapy is used to lower lead levels in the body, most of the lead that is removed comes from the bone and soft tissues back into the blood and then is excreted in urine. Unfortunately, lead in the brain is not removable by chelation. Similarly, lead in the brain is not affected by the equilibrium processes discussed in the preceding paragraphs. Once lead enters the brain, it does not move out again.[8] This situation can have serious consequences since the brain is an organ that is very sensitive to lead and in which significant effects can occur.

Estimates of the amount of lead in bone and soft tissues at various ages have been obtained from autopsies of people who died at different ages. Lead in soft tissues appears to increase with increasing age up to about 40-50 years, after which it begins to decline. Lead in bone follows a similar pattern, although the point at which it shifts from an increase to a decrease is at about 50-60 years.[9] Lead in blood is highest at about two to three years and begins to decrease rapidly after about age five or six. After about age 17, a gradual increase begins that continues to about age 50 to 60 after which a slight decrease occurs.[10]

Lead does not stay in the blood for a long time period. Red cells are constantly forming and, after about 120 days, breaking down. When they break down, much of the lead that was in the cells may be excreted in urine.[11] For this reason, measuring lead in the blood is a good measure of recent lead exposure but is not a valid measure of other exposures or of the total body burden of lead. Some other, very small amounts of lead are also "excreted" in sweat, nails, breast milk, and hair.

Although there may be a great deal of variability, lead stays in soft tissues on the average for about 30 days.[12] The largest amount of lead in soft tissues is found in the aorta (the main artery carrying blood from the heart to the rest of the body), then the kidney and liver, and then the brain.

A final source of lead movement that needs to be considered is that between a mother and her fetus or breast-fed child. Lead in a mother's blood can pass directly into the blood of her unborn child with only a slight decrease in amount as it crosses the placenta. The amount of lead in the mother's blood will become relatively more concentrated in the fetus when considered on the basis of their respective body weights. In addition, the rapidly developing fetal system is very sensitive to lead.

After the birth of the child, a nursing mother can pass additional amounts of lead on to her child in her breast milk. It has been recommended that less than 100 micrograms of lead a day is an acceptable level for an infant.[13] Since an infant consumes up to about 80 decilitres of breast milk a day, the lead level would have to be less than 1.25 μg per decilitre of milk to keep the daily total below 100 μg for a totally breast-fed infant. A recent Canadian study of lead levels in 210 nursing mothers found an arithmetic average lead level in breast milk of only about 0.01 μg/dl.[14] This actual level is considerably smaller than the allowable limit of 1.25 μg/dl. It appears then that breast milk does not seem to be a major problem. This conclusion is further supported by the fact that, in general, blood-lead levels in children one year or less are lower then they are in slightly older children. However, since an infant is quite sensitive to lead from any source and since many very young infants depend on breast milk for their entire food supply, a nursing mother who feels that she might have had extensive lead exposure, might wish to ask her doctor to have her milk tested for lead. Women who are pregnant or who plan on getting pregnant would be wise to follow as many lifestyle risk reducers as possible.

Because of the complicated equilibrium processes, the fact that lead in the brain cannot be removed, and the fact that lead in bone can come out of storage in later years to again enter the circulatory system, the best protection from lead-caused health problems is to keep lead from getting into the body in the first place.

Chapter 9

Your Blood-lead Level

How to Interpret Your Blood-lead Test

The results of your blood-lead test will probably be reported to you as so many micromoles of lead per litre of blood (μmol/L), or possibly as so many micrograms of lead per decilitre of blood (μg/dl). If it is reported in μg/dl, you do not need to make any conversion of this value to be able to compare it to the risks and levels used in this book. If it is reported in μmol/L, ask your doctor what the value is in μg/dl or make you own conversion to μg/dl.

For blood-lead tests, the conversion factor from μmol/L to μg/dl is 20.7. For example, if the results of your test are reported as 0.61 μmol/L, simply multiply 0.61 times 20.7 to find out what your level is in μg/dl. In this example, the blood-lead level is 12.6 μg/dl.

A blood-lead test measures the amount of lead in your blood at a specific time and represents the amount of lead recently absorbed and the amount absorbed in the past that is still in the blood or that has moved back into the blood from some storage site in the body. This internal movement of lead is described in the previous chapter.

A blood-lead test measures this combined amount of lead and should not be confused with the amount of lead absorbed in a day. Since absorption and blood-lead level measure two different things, the numerical values that indicate serious concern are different for these two measures. As mentioned in the account in Chapter 1 on lead absorption in the aristocratic classes of ancient Roman society, 250 micrograms (μg) of lead absorbed daily will cause lead poisoning. This 250 μg/day poisoning value can be contrasted to the blood-lead level of 100 μg/dl that is sufficient to cause death. All of the information in the rest of this chapter will focus on blood-lead levels (μg/dl) rather than on absorption rates (μg/day).

Once you have determined what your or your child's current blood-lead level is in μg/dl, you can consult Tables 6.1 and 6.2 to see what possible health effects or biochemical changes might be occurring at that level. There are several things that you should keep in mind when you compare your results with the values shown in these tables. **The first of these is that the tables show the lowest blood-lead level at which the effects have been shown to occur in some, but not all people.**

Table 9.1

Medical and Lifestyle Actions at Different Blood-lead Levels

Blood-lead Level (μg/dl)	Medical and Lifestyle Actions
	Adult Males and Non-childbearing Females
over 30	Blood-lead level retested and possible investigation of exposure sources; important to follow lifestyle risk reducers carefully.
20-30	Possible blood-lead retest and exposure investigation; wise to adopt lifestyle risk reducers.
under 20	No action; follow risk reducers, if desired.
	Adult Childbearing Females and Children from 6 to 18 Years
over 20	Blood-lead level retested and investigation of exposure sources; lifestyle habits should be changed; if not yet pregnant try to lower the level using risk reducers before getting pregnant.
under 20	No blood-lead retest unless requested; follow risk reducers to get the blood-lead level below 15 μg/dl if possible.
	Children Under 6 Years
over 25	Blood-lead level retested, exposure sources investigated, possibly chelation treatment used; lifestyle risk reducers should be very carefully followed.
20-25	Usually blood-lead level retested and exposure sources investigated; important to follow lifestyle risk reducers carefully.
15-20	No blood-lead retest unless requested; follow risk reducers to get the blood-lead level below 15 μg/dl if possible.
0-15	Follow risk reducers if desired.
below 10	Probably no action is needed.

The second point to consider is that there is always a margin of error. The result that you receive on your blood-lead test is the best measure you have of the actual amount of lead in your blood at the time the test was made, but it is not an absolute "truth." If your level seems high and you are concerned, you might want to get a second test to see if any unintentional measurement errors occurred in the first test. blood-lead testing is a very sensitive procedure and a repeat test may give you a more accurate picture of your personal level.

A third point to consider is the role of individual differences. **Depending on a number of factors, a particular individual may show effects at blood-lead levels higher or lower than those in the tables.** Some of the factors that determine these individual differences are hereditary sensitivity to lead, amount of lead in the brain from past exposures, amount of calcium and iron in the diet, amount of lead in storage in the body that is not circulating in the blood, whether or not the present blood-lead level is being affected by a recent large increase or decrease in absorbed lead, and other factors.

After taking all these factors into account, the rough guidelines in Table 9.1 can be used to help you decide if there is any further action you should take. These guidelines are based on the discussion of "Risks and Levels" in Chapter 6. When the phrase "childbearing" or "non-childbearing" is used, it refers to women who are (or are not) pregnant and to those who plan (or don't plan) to have children in the future.

Estimated Blood-lead Levels

Figure 28, "Estimated Blood-lead Levels by Sources," is a summary of the information in Appendix G on the estimated amounts of lead in blood that are due to different sources. These estimates are shown separately for children and adults and for people living in rural, urban, or urban lead industry (ULI) areas. Note that the ULI category refers only to residences located quite close (about 200 metres) to a lead industry.

A detailed description of the basis for each of these figures can be found in Appendix G. If you have specific information on lead levels for any of these sources in your immediate environment, you might wish to use the formulas in Appendix G to develop estimates that would be more specific for your area.

Note that the only sources considered in these estimates are food, dust, air, and water. While these four sources represent the major contributors to blood lead for most people, there are other sources (see Chapters 3 and 7) that can also make contributions. The most important of these other sources pertain primarily to adults and include occupational sources, hobbies, and smoking.

Perhaps the most striking part of this figure is the large amount of blood

Figure 28

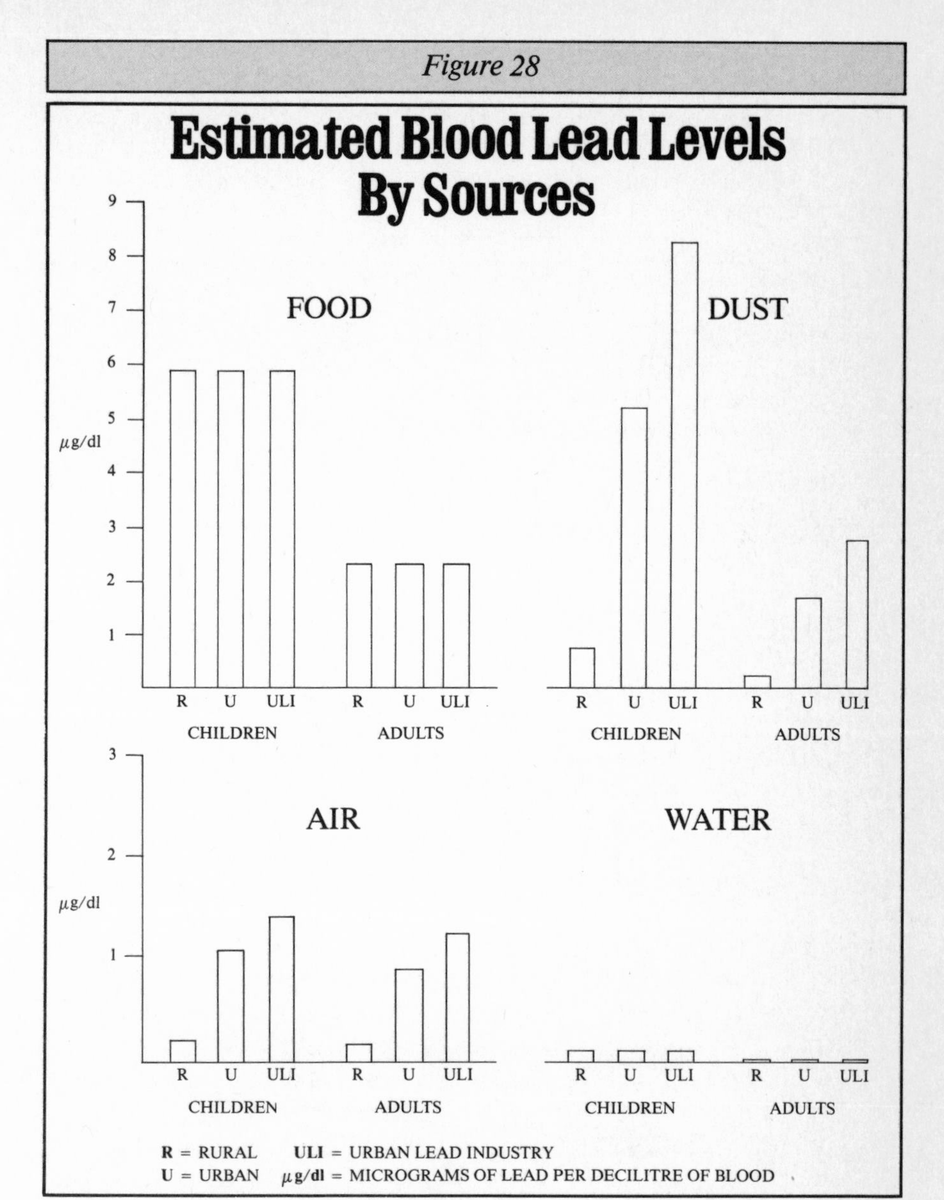

Estimated amounts of lead in blood that come from lead in food, dust, air, and water for two-year old children and adults living in rural, urban, and urban lead industry areas. Note that adults may have additional amounts of lead in their blood due to occupational exposure (variable amounts), consumption of alcohol, (adds about one μg/dl), and smoking (adds about two μg/dl).

Source: *Appendix B.*

lead that comes from food and the very small amount from water. The food lead contribution is more than twice as large for children as for adults due primarily to the fact that children absorb much more of the lead in the gastrointestinal tract than do adults. For both food and water, there is no difference between rural, urban, and urban lead industry areas for a particular age group.

Figure 29

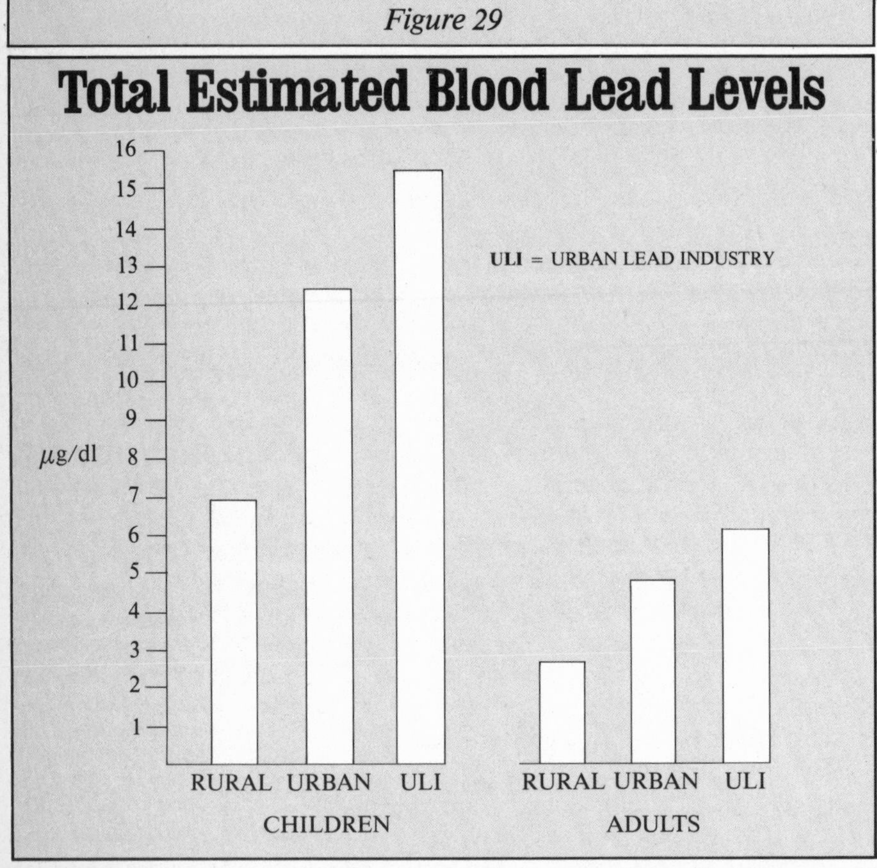

Total estimated amount of lead in blood for two-year old children and adults living in rural, urban, and urban lead industry areas. Note that adults may have additional amounts due to occupational exposure, consumption of alcohol and smoking (see preceding figure).
Source: *Appendix B.*

Dust and air contributions to blood-lead levels vary with the place of residence. In both cases, there are progressively larger contributions as you move from rural to urban to urban lead industry areas. For air lead contributions, there is not a large difference between the values for children and adults. However, for dust lead contributions, both age and place of residence make a considerable difference. These variations in dust lead contributions to blood-lead levels can be reduced in the case of a particular child or adult by carefully following the dustbusting suggestions in Chapter 3.

The amounts separated by sources in Figure 28 can be combined for each of the age and residential groups to produce a total estimated blood-lead level for each group. These totals are shown in Figure 29, which very clearly shows the greater amount of lead in the blood of children than in adults and the increasing amounts in both age groups in urban and urban lead industry areas as compared to rural areas.

The bars in Figure 29 represent the estimated average blood-lead level for each of the six groups. This average value, however, does not describe the number of individuals in the group that may be expected to have blood-lead levels above or below this point. For example, although the average value for urban children is about 12 μg/dl, roughly half the children will have levels above this number and roughly half will have lower levels. Some of these children will have values considerably higher (or lower) than the group average. This fact should be kept in mind when these averages are compared to the various levels representing "elevated," "intervention," and "biochemical risk."

All of the average values on this graph are below the "elevated" level for children of 25 μg/dl and the "intervention" level for children of 20 μg/dl. The average values for the urban child (about 12) and the urban lead industry child (about 15) are higher than the beginning "biochemical-risk" level of 10 to 15 μg/dl. The urban child average falls right in the middle of this range. All of the adult averages and the rural child average are below the "biochemical-risk" level, which indicates that except for individuals in these groups with values well above the average, most people in these categories do not appear to be experiencing lead health problems. The one exception to this generalization is the possibility of blood pressure effects in adult males at levels as low as 5 μg/dl.

The group with the highest average value is urban children living near lead industry point sources. Since the average child in this group is estimated to have a blood-lead level high enough to be causing biochemical changes, many children in this situation will have even higher levels with the result that most of these children are probably experiencing some kind of health risk due to their lead exposure. This situation is one of the reasons why we are recommending substantial revisions in lead control strategies.

Actual Blood-lead Levels

When actual blood-lead levels are determined for individuals or groups, there is no way to know where the lead comes from. The only information on sources comes from estimates based on the kinds of formulas used to derive the numbers shown in Figure 28. If the total estimates (based on the sum of the different source estimates) agrees with actual blood-lead measurements, then there is a good possibility that the source breakdown values may be approximately correct. Information about the relative contribution of each source is valuable in determining how to reduce lead levels in the blood.

There is only a limited amount of actual data available on blood-lead levels in large groups of people in Canada. There are almost no recent, reliable data on actual levels in Canadian adults except for one 1981 study in Alberta on 338 blood donors who had an average blood-lead level of 10.1 μg/dl.[1] There is, however, a 1984 Canadian survey that measured blood-lead levels in 1269 children living in rural, suburban, and urban areas in Ontario. This study is referred to here as the Tri-ministry Study since the data were gathered under the combined auspices of the Ontario Ministries of the Environment, Health, and Labour.

The largest blood-lead study (NHANES-II, for the second National Health and Nutrition Examination Survey) took place in the United States from 1976 to 1980. Over 10,000 blood lead tests were performed on a cross-section of the population from six months to 74 years old. This study found that blood-lead levels ranged from two to 66 μg/dl with a median value (i.e., half the people had higher and half, lower lead levels) of 13 μg/dl.[2] Some of the more striking between-group comparisons in this study included higher blood-lead levels for:

- urban than rural groups (15 to 13 μg/dl);
- blacks than whites (16 to 14 μg/dl);
- men than women (16 to 12 μg/dl);
- low income than higher income groups (20 to 14 μg/dl in children under age six); and
- children under age six than older age groups (16 to 12.5 and 14 μg/dl).

In addition, over the four years that blood-lead samples were taken, lead levels declined significantly and in a lock-step manner with the declining use of lead in gasoline (see Figure 30 in Chapter 11). These general downward trends are continuing so that an approximate estimate of blood-lead levels in urban U.S. children under age six is about 10 μg/dl in 1985.[3]

Results of the Ontario Tri-ministry Study are shown in Table 9.2. The overall average level was 10.4 μg/dl with higher levels in urban than suburban and in suburban than rural areas. Note that while these 1984 values for Canadian children are lower than the U.S. values obtained in 1976 to 1980, they are not significantly different from the 1985 U.S. estimate of about 10 μg/dl.

The overall average and the averages for rural and suburban children in the Tri-ministry Study are very near to or below the "biochemical-risk" level of 10 to 15 μg/dl. The urban average of 12.0 μg/dl falls within this risk range. These average values, except perhaps the urban average, present a fairly reassuring picture of the Canadian blood-lead situation. However, the average values do not give a complete picture.

Table 9.2

Percent of Ontario Children Aged Six and Under in Various Blood-lead Categories by Place of Residence

Residence	Blood-lead (μg/dl)				Geometric Average (μg/dl)
	under 10	**10—19**	**over 20**	**over 25**	
Urban	29.1%	65.3%	5.6%	1.8%	12.0
Suburban	51.9%	44.4%	3.8%	1.6%	10.0
Rural	63.2%	33.3%	3.5%	0.8%	8.9
Total	48.1%	47.6%	4.2%	1.4%	10.4

Note: The totals may not sum to 100% due to rounding differences. The total number of children tested was 1269 and did not include any children from urban "hotspots."

Source: *C. Duncan, R. A. Kusiak, J. O'Heany, L. F. Smith, L. Spielberg, and J. Smith, "Blood Lead and Associated Risk Factors in Ontario Children, 1984," Ontario Ministries of Environment, Health, and Labour, 1985.*

When the percentages of children above 20 and above 25 μg/dl are considered, this reassuring picture begins to fade. There are 1.4% of the children over 25 μg/dl, 4.2% over 20 μg/dl, and 51.8% (47.6% + 4.2%) over 10 μg/dl. Using these Ontario percentages as the best available estimate of the distribution of blood-lead levels in Canadian children,[4] it is possible to estimate the number of children of the same age (under seven years) in the entire country that might fall into these categories.

Applying these percentages to the total Canadian child population of this age (2,496,750 in 1981) produces some disturbing predictions. **For children under the age of seven in Canada, about 35,000 children may be above the "elevated" blood-lead level; about 100,000, above the suggested "intervention" level; and about 1,290,000 (or just over half of all children this age), above the lower limit of the "biochemical-risk" level. These large numbers indicate that many of our children may not be developing their full potential and may be at risk of lead health effects.**

Table 9.3 compares the estimated blood-lead levels in Figure 29 with the actual levels obtained in the Tri-ministry Study, in recent local measurements around a lead industry in Toronto, and in the 1981 Alberta blood-donor study. This comparison is one way of testing the validity of the estimated values. For the three children's groups, the estimates and actual values are highly similar. For both adult groups where comparisons can be made, the estimates are considerably lower than the actual values. The reason for these large adult differences may be the fact that the estimates do not take into consideration lead from occupational exposure, smoking, alcohol consumption, or hobbies.

Table 9.3

Comparison of Estimated and Actual Blood-lead Levels

Group	Estimated	Actual	Actual-Estimated
Children-rural	6.98	8.91	+1.93
-urban	12.32	12.02	-0.30
-ULI	15.60	15.00	-0.60
Adults -rural	2.63	NA	NA
-urban	4.84	10.10	+5.26
-ULI	6.10	11.60	+5.50

ULI = urban lead industry

Sources: *Appendix G; for the actual adult urban value, J. I. Cheng, "A blood lead survey of Southern Alberta," paper presented to Can. Public Health Assoc. Conference, Saskatoon, June 26, 1981; and for the actual adult ULI value, E. Ellis, Toronto Department of Public Health, personal communication, May, 1986.*

It is also interesting to compare the actual Canadian and U.S. values with fairly recent data from remote areas of the world. Adult blood-lead levels from Yanomamo natives living in a very remote region of the Amazon basin had a maximum level of approximately 3 μg/dl.[5] Children in another fairly remote area (Papua New Guinea), had an average blood-lead level of about 4 to 5 μg/dl with a maximum level of only 13 μg/dl.[6] On a island (Isle of Aran) off the west coast of Ireland where there are no motor vehicles, average values for adult men were 5.6 μg/dl and for adult women were 4.7 μg/dl.[7] All these values are considerably lower than Canadian and U.S. values. The effect of our more highly industrialized society and the use of leaded gasoline in our vehicles is clearly reflected in the amount of lead in the blood of both adults and children.

Chapter 10

The Leaded Gasoline Story

A Lead Tiger

Back in the late 1960s when Esso was putting a "tiger in your tank," just about the only kind of gasoline you could buy was leaded. At that time, approximately 20% of all lead used was for the manufacture of gasoline lead additives. This percentage has since shrunk to about 6% as countries around the world have gradually phased down the lead content of gasoline and as increasing numbers of cars, especially in North America, require unleaded gasoline.

Lead additives act like a "tiger in your tank" because they boost the octane level of the gasoline. The octane level or "number" of a gasoline indicates the quality of that fuel based on its antiknock properties. The "boost" in octane number provided by lead additives acts to raise engine compression and, therefore, speed by avoiding problems associated with pre-mature combustion of the air-fuel mixture (i.e., "knocking") in an engine cyclinder.

Although the principal reason that lead is added to gasoline is to boost the octane level, lead also acts as a valve lubricant by forming a protective coating on the exhaust-valve seat. If lead is not present, abrasive and adhesive wear on valve seats can cause a problem called "valve seat recession." This problem can require major engine repairs. Fortunately, valve seat recession is only a problem in a very small number of vehicles. Those vehicles that are susceptible are some foreign or pre-1971 North American vehicles with unhardened valve seats, or without valve-seat inserts, and only if engines in those vehicles experience very high "rpm's" (revolutions per minute) for extended periods of time (tens to hundreds of hours).[1] Engines will experience high rpm's when operated at very high speeds (i.e., greater than 70 mph or 113 km/h) or while carrying very heavy loads. All North American vehicles built after 1971 have hardened valve seats or other features designed to minimize valve wear, and do not need lead as a valve lubricant. Most foreign built vehicles have hardened valve seats or similar features but, to be certain of a particular foreign vehicle's characteristics, consult the owner's manual or automobile dealer.

It is clear that the vast majority of vehicles designed to use leaded gasoline will operate perfectly well on unleaded gasoline. Even in those vehicles that actually require lead, the amount of lead needed is quite small. The recent lowering of the leaded gasoline standard in the U.S. to the very low level of 0.026 grams of lead per litre of gasoline (0.026 g/L) was chosen to ensure the prevention, with an adequate safety margin, of valve seat recession in vehicles susceptible to it.[2] The issue of gasoline lead standards in Canada is discussed more fully in Chapter 13.

Maintenance and Fuel Economy

Although the general public is not usually aware of the fact, that "lead tiger" is doing more to car engines than boosting octane, eliminating "knocking," and lubricating valves with unhardened valve seats. It is also causing corrosion and rusting inside the engine and exhaust system[3] and, as already noted, is the number one contributor to environmental lead contamination.

The lead in the gasoline is not "used up" in the combustion process. It forms new lead compounds, and about 75% of the lead originally in the gasoline mixture goes out the tailpipe after combustion.[4] About 10% of the lead ends up in the engine oil where it is eventually removed during oil changes. Another 15% forms deposits on the interior surface of the engine and exhaust system. Some of this deposited lead eventually flakes off in large pieces and leaves the engine with the exhaust gases.

The corrosive lead deposits in the engine and exhaust system cause higher maintenance costs due to the need for more frequent spark plug changes, oil changes, and exhaust system repairs than with unleaded gasoline. In addition, there is reduced fuel economy with leaded compared to unleaded gasoline.

When all the factors are taken into account, a car that was originally built to use leaded gasoline will cost roughly the same amount to run whether it uses unleaded or leaded gasoline as long as unleaded gasoline costs more at the pump. If there is no at-pump price difference, driving an older car on leaded would be wasting money.

If a car that was built to use unleaded gasoline is misfuelled with leaded gasoline, the car's owner definitely loses money. Figure 8 in Chapter 2 shows this situation graphically. The car owner who misfuels destroys the car's pollution control devices (e.g., the catalytic converter) and loses about $400 every 80,000 kilometres in additional maintenance costs and reduced fuel economy.[5] The cost breakdown of these losses can be found in Appendix H.

Lead and Catalytic Converters

Unleaded gasoline was first introduced in the United States back in the early 1970s because of its indirect relationship to the control of other automobile pollutants. At that time, all new cars in the U.S. were required to have catalytic converters in order to meet federal standards for the amount of hydrocarbons and carbon monoxide that could be emitted in car exhaust. Since these catalytic converters could not tolerate the lead in leaded gasoline, another regulation was passed requiring that unleaded fuel be made available. At about the same time, Canada and Mexico also introduced unleaded fuel, initially for the sake of American tourists driving the new cars.[6]

By the late 1970s, the original "two-way" catalytic converters had been replaced by "three-way" catalytic converters that could also control emissions of nitrogen oxides. The Canadian government has formulated emission regulations (effective September, 1988 for 1988 model cars) for hydrocarbons, carbon monoxide and nitrogen oxides at levels low enough to require high quality pollution-control devices. It will be up to the car manufacturers to decide how these new emission regulations will be met. The best available technology to meet these limits is the three-way, lead-intolerant, catalytic converter.

Alternatives to Lead

Instead of using lead additives, the required octane level in unleaded gasoline is produced by slightly more expensive refining processes. The added refinery expense is incurred either as an increase in regular operating expenses or by capital investment in additional refining equipment. Some of these refining processes involve an increase in the aromatics content of the gasoline and some do not. Aromatics such as benzene (a well-recognized carcinogen), toluene, and xylene are commonly found in the crude oil used to produce all types of gasoline. Refineries can use the aromatics either as octane boosters in their gasoline or as components of other products. If a refinery uses its aromatics for non-gasoline related products, it will use other refining processes, such as isomerization or the addition of alcohols, to boost octane levels. The option or combination of options chosen by individual refineries to boost octane levels depends upon site-specific variables such as the type of crude oil being processed, the range of products typically produced, and the equipment in place.[7]

As a result of the wide variety of refining technologies in place across the country, reductions in the lead content of gasoline and the increased use of unleaded gasoline will mean that there will be more aromatics in all types of gasoline. The presence of the aromatic benzene in gasoline is a special environmental and health concern. However, even though it is estimated that the benzene content of the gasoline pool will increase as lead is phased out of gasoline, experts in the U.S. have predicted that overall emissions of benzene will actually decrease. This benzene-reduction prediction is based on the fact that the newer vehicles designed to use unleaded fuel have better pollution control equipment and that misfuelling is expected to decrease with a lower lead standard.[8]

Other additives, such as MMT or alcohols, are also able to boost the octane level of gasoline. MMT is based on manganese, another metal that is hazardous to health. Using MMT to boost octane can save money for refineries compared to other octane-boosting alternatives which involve increased operating costs or capital investment in new refining technologies. However, this financial advantage to the industry must be balanced against the potentially large problems that could arise from the use and environmental dispersal of MMT.

Alcohols, such as methanol, are not actually alternative additives but rather gasoline *extenders* that can boost the octane level of gasoline as effectively as lead without any of the associated environmental hazards. The use of methanol as an octane booster not only reduces lead emissions but also reduces emissions of aromatics. The early problems with methanol in gasoline, including difficult cold-starting and separation from gasoline in the presence of water, have now largely been resolved.[9] The amount of methanol produced from natural gas at a large plant in Edmonton, Alberta is apparently large enough to satisfy the demand, as an extender, for the entire Canadian gasoline pool.[10] In fact, straight methanol is used as a fuel replacing gasoline in specially-designed engines. Such use may eventually replace gasoline on a large scale since current technology used in the storage, transportation, and consumption of gasoline can be applied to methanol relatively easily.[11]

From Tank to Blood

The relationship between the use of leaded gasoline and the amount of lead in people's blood became startlingly clear when U.S. scientists compared the drop in the production and use of lead additives for gasoline in the late 1970s with a corresponding and unexpected drop in blood-lead levels across the country.

The Second National Health and Nutrition Examination Survey (NHANES-II) measured blood-lead levels (and other information) in a large sample of U.S. residents from 1976 to 1980. Figure 30 shows the strength of the relationship that was found between blood lead and gasoline lead. Blood-lead levels follow the overall downward trend in leaded gasoline production as well as the seasonal variations in emissions of gasoline lead (which increase in summer when more gasoline is used). Statistical testing revealed that this relationship held true even when the possible effects of other variables related to blood lead (e.g., exposure to other sources of lead, nutrition, income) were taken into account. Scientists from Ethyl Corporation and Dupont, both large producers of gasoline lead additives, drew different conclusions from the NHANES-II data.[12] These conclusions have since been discredited by a Review Group[13] that supported the evidence presented by the Centers for Disease Control in Atlanta and by the United States Environmental Protection Agency showing the strong relationship between blood lead and gasoline lead.

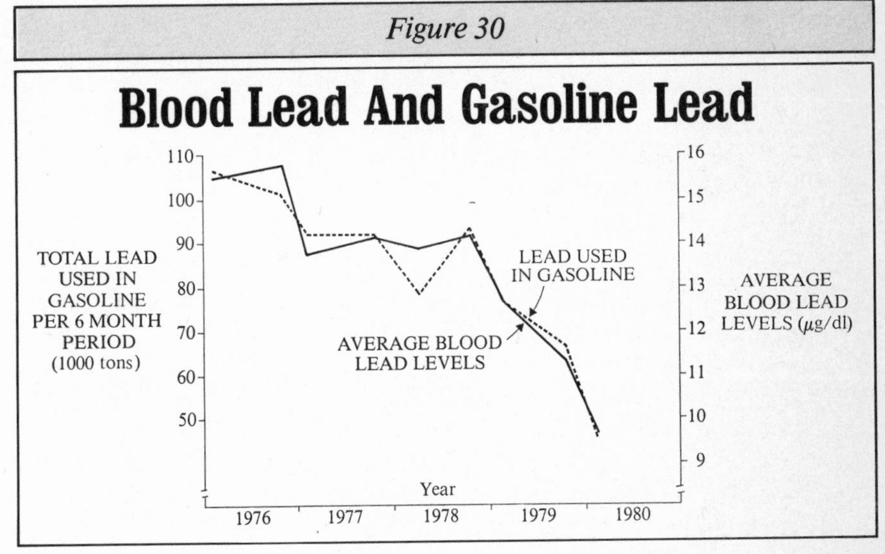

Trends in the average blood lead levels of people living in the United States (February 1976 - February 1980) and lead used in gasoline production.
Note: *One ton equals about 0.98 metric tonnes.*
Source: *Adapted from U.S. Environmental Protection Agency, "Costs and Benefits of Reducing Lead in Gasoline," (EPA-230-05-85-006), 1985, p. III-8.*

It is estimated from NHANES-II that, during the 1970s, lead from gasoline accounted for at least half of the lead in the blood of U.S. residents.[14] A more recent Canadian study[15] places this value for Canadians at the somewhat lower level of 30-40%. Gasoline lead may also be contributing further amounts to blood lead by way of indirect sources that are very difficult to determine. Such further contributions are probable since gasoline lead emissions are responsible for 60-80% of environmental lead contamination.

It has taken almost 15 years of extensive research to come to some strong conclusions about the hazards to human health of environmental lead contamination and the major role of gasoline lead in this contamination. These conclusions have been reached in an atmosphere of scientific controversy since throughout this time, powerful corporations with economic interests vested in lead have funded research that appears to refute these claims. In addition, large corporations are able to exert considerable political influence through exceptionally well funded, professional lobbying efforts. Such pressures in the political arena can be the basis for slow and compromised development of regulatory responses to known environmental problems.[16] This situation may be true particularly in Canada where the large lead industry probably has something to do with the fact that the 1986 legal limit for lead in gasoline is the most lenient in the industrialized world.

Denials in the 1920s[17]

The effectiveness of industrial lobbies on governmental regulation of lead in gasoline is not new. The addition of lead to gasoline in the 1920s set off a storm of controversy involving similar issues to those faced by the public, scientists, industry and regulators in the 1970s and today. At that time, several cases of industrial lead poisoning, including deaths, occurred during the early production of tetraethyl lead (TEL), the first chemical formulation of gasoline lead additives. Independent scientists and industrial hygienists questioned the safety of using a known toxin as a gasoline additive, citing the poisonings and pointing out that the lead dust in the exhaust would result in an unknown public health risk.

Despite such concerns, leaded gasoline went on the market before possible public health risks had been assessed. In response to demands for research into these possible risks, an agreement was made in 1923 between the U.S. government and General Motors (GM) Research Corporation whereby GM would pay for a study done by the U.S. Bureau of Mines. GM managed to get clauses put into the agreement that would bar press and progress reports during the study so public anxiety would not be aroused. The word "lead" was also omitted from correspondence about the study and the trade name "ethyl" was used instead. By the end of the study Ethyl Corporation, which had been formed by GM and Dupont to produce "ethyl" gasoline, had negotiated exclusive rights for comment, criticism, and approval of any of the research generated by the study before it was released. Under such circumstances, it was not surprising that leaded gasoline was given a clean bill of health.

In 1924, at the same time that the U.S. Bureau of Mines began to assure the American public of the safety of leaded gasoline, a disastrous series of poisonings occurred in the Standard Oil Company's experimental laboratories. Over the space of four days, five out of 49 workers died, and 35 others experienced symptoms of severe lead poisoning. In the face of this industrial disaster and other known dangers of lead, eminent scientists, public health experts, and labour activists across the country attacked the industry-funded study that had absolved leaded gasoline, claiming that it was inadequate and biased.

The Surgeon General of the U.S. Public Health Service convened a conference in May of 1925 to attempt to resolve the controversy. The conference brought together many different points of view but resulted in little concrete information on which to base a decision. There was at that time a general lack of information about the possible long-term public health effects of leaded gasoline. This situation generated an attitude among many public health experts that it was unfair to ban leaded gasoline without any definitive proof of harm to the public at large. More long-term studies were recommended to resolve the debate.

A "Blue Ribbon Committee" was then set up to conduct another study and instructed to provide answers quickly. Under such circumstances the committee could conduct only a very limited study (which was, in fact, the opposite of what had been recommended). After a few months the committee found no reason to ban leaded gasoline. However, it warned that

> ...it remains possible that if the use of leaded gasoline becomes widespread (sic) conditions may arise very different from those studied by us which would render its use more of a hazard than would appear to be the case from this investigation.[18]

The committee recognized the limited ability of such a short-term study to detect long-term effects and advised that onging studies by the government were essential.

> ...the committee feels that the investigation begun under their direction must not be allowed to lapse...It should be possible to follow closely the outcome of a more extended use of this fuel and to determine whether or not it may constitute a menace to the health of the general public after prolonged use or other conditions not now foreseen... The vast increase in the number of automobiles throughout the country makes the study of all such questions a matter of real importance from the standpoint of public health and the committee urges strongly that a suitable appropriation be requested from Congress for the continuance of these investigations under the supervision of the Surgeon General of the Public Health Service.[19]

These recommendations were never heeded. There were no long-term studies and further investigations were conducted by the industry itself.

This summary of the controversy in the 1920s illustrates how difficult, if not impossible, it is for "science" to resolve an issue that is more than just a scientific debate. The discovery of this powerful new antiknock compound was heralded by representatives of Ethyl Corporation as a "gift of God"[20] and linked to the industrial progress of America. The cases of lead poisoning in workers handling leaded gasoline were blamed on worker carelessness and failure to follow instructions. Objections to the widespread use of leaded gasoline on the basis of risks as yet unproven by science were dismissed, noting that any great innovation involved some risk.

Scientific investigations throughout the 1920s could not say that the widespread use of leaded gasoline would be safe or hazardous to the public. Suggestions of uncertainty, of risks, and of the need for much more extensive research to resolve unforseeable but potentially significant public health risks were largely ignored, possibly because such suggestions did not fit comfortably into the economy and society of the 1920s, a society that was increasingly built around a firm belief in industrial progress geared to the automobile.

The Evidence Today

As we have seen, it is now known that lead in gasoline contributes 60-80% to the lead contamination of the environment, and about 40% to the amount of lead in blood. Adverse health effects of lead have been shown to occur at progressively lower blood-lead levels as the information base has grown. The best available estimates in 1985 of Canadian blood-lead levels are cause for concern since they indicate that over 35,000 children under seven years may have "elevated" blood-lead levels and about half of all Canadian children may have levels at or above the 10 to 15 μg/dl biochemical risk level.[21] In addition, the link between lead and blood pressure effects in adult males is becoming stronger as research continues.

The Canadian regulatory response to the problem of lead in gasoline is about five years behind the United States, which has, as of January 1, 1986, a gasoline lead standard that is 30 times lower than Canada's.[22] The U.S. decision to set a reduced level of 0.026 g/L was based on economic analyses that included predictions of significant cost savings for children's health care and remedial education and the fact that consumers will save millions of dollars in reduced maintenance costs and improved fuel economy.[23] In addition, if the evidence for lead effects on blood pressure is shown to be reliable, these anticipated health care savings in the U.S. would increase further by over 38 billion dollars between 1985 and 1992. (Appendix H contains a summary of the U.S. cost-benefit analysis). Although Canada has not undertaken a comparable economic analysis, it is highly likely that the same kinds of economic arguments apply to the Canadian situation since average blood-lead levels are similar in both countries.

In summary, lead additives are at an unnecessarily high level in Canadian gasoline and are causing the consumer to spend more on maintenance and fuel than if the level of lead were sharply reduced. In fact, the vast majority of vehicles that were originally built to use leaded gasoline can operate more efficiently without lead additives in their gasoline. For the

Motorist's Questions

1. **My car was built to use leaded gasoline. Can I switch to unleaded?**
   Yes. The only cars originally built to use leaded gasoline that can't switch to unleaded are those built before 1971 (North American; for foreign car's consult a dealer or owner's manual) that have unhardened valve seats or valve-seat inserts and that are operated at high rpm's (high speeds or heavy loads) for extended time periods.

2. **What about other engines like my lawn mower or my snow mobile? Can they run on unleaded too?**
   Yes. Small gasoline powered engines such as lawn mowers, snow mobiles, garden tillers, and snow blowers can all operate on leaded or unleaded. Manufacturer's usually recommend unleaded for better engine performance and fewer maintenance repairs.

3. **What about my boat engine, can it use unleaded too?**
   Inboard motors are like car engines and can use unleaded gasoline just as easily as can the majority of cars. Outboard motors, which are usually two-stroke engines with the fuel pre-mixed with a special kind of oil, can all use unleaded gasoline except for the high-output, two-stroke engines (125 horsepower or higher) which need the lead additives.

4. **Are there safe alternatives to lead for the small number of vehicles and engines that still need lead as a valve lubricant?**
   Not at present although experimental work in the U.S. has shown phosphorous compounds to be adequate valve-seat lubricants. Such compounds can damage catalytic converters and would therefore have to be used in non-catalyst equipped engines (which applies to most of the engines that require lead for valve durability) or as a separate additive.

5. **Why is leaded gasoline cheaper if they have to *add* lead to it?**
   Adding lead-containing chemicals is a cheap way for refineries to boost the octane level of gasoline. The octane level of unleaded gasoline is achieved by slightly different, and slightly more expensive refining processes. The difference in refining costs is smaller than the price difference at the pump. Since the regular leaded gasoline price is the "hook" to get you to drive into the station, the inflated price difference is a marketing strategy.

6. **Why doesn't the government just equalize the price of leaded and unleaded gasoline to stop misfuelling?**
   If the public wanted this done and was vocal enough about it, it would happen.

very few vehicles that actually require lead, the amount of lead in gasoline only needs to be at the very low level of 0.026 g/L.

In the 50 years since lead was first added to gasoline, we have been participating in an uncontrolled experiment on the effects of widespread lead exposure on the environment and human health. We have contaminated the environment with lead so effectively that current *average* blood-lead levels in children are at the level where known changes begin to occur in the human body. These levels are so high that there is almost no safety margin below the level at which obvious health problems occur. Compared to the *potential* risks for the general public posed by other pollutants of major concern today, (such as PCBs, dioxins, and pesticides in food and drinking water supplies), lead is a much more important pollutant to bring under immediate control since it is causing *actual* health problems now to a significant number of people.

Chapter 11

Balancing Dollars

The Lead Industry

A wide variety of companies around the world produce lead and lead-bearing products, making lead the fifth most-used metal at a total annual production of over 5 million tonnes. Lead production includes the mining, milling, and smelting of lead-bearing ores to produce pure lead ingots (primary production) and the recovery of lead from lead-bearing products such as used car batteries and cable sheathing (secondary production). Canada is the third largest producer of lead in the western world, after the U.S. and Australia.

Although there are many different uses for lead, the market for lead-bearing products is dominated by the manufacture and use of automobiles. For example, in Canada, battery manufacturing uses over 50% of all lead produced[1] and will likely continue to dominate the lead market since, despite many years of research, there does not at present seem to be any reasonable alternative to the use of lead in automobile batteries. Additional automobile-related lead products include lead as a gasoline additive, solder for body work when cars are built or repaired, and lead-containing paints, such as the bright yellow on some service vehicles and the yellow and white lines on streets and highways.

Our modern society has grown to depend on lead for a number of functions. Some of these, such as medical x-ray shielding, are presently irreplaceable. Others, such as gasoline lead additives, can be easily replaced. We derive benefits from lead in the form of jobs, useful products, exports, and from all of these, federal and provincial tax revenues.

The recovery of lead during secondary production also provides environmental benefit by preventing large amounts of lead from entering the environment when lead-containing products wear out. The most common of these products are car batteries, which typically last three to four years.

At present a large proportion of used batteries are recovered by the secondary lead industry. Although this type of recovery operation is an environmental benefit on a wide scale, it can be a significant local environmental hazard unless emissions from the secondary smelter are properly controlled. Since most old batteries are found in cities, secondary lead smelters in Canada tend to locate in urban areas sometimes causing severe local lead contamination, which adds to already high urban lead exposure levels.

Lead from old batteries is not recovered, however, for the sake of providing environmental benefits to society. Lead recovery is done to make money. If the price of lead drops, secondary production slows and collection of scrap is limited to nearby, high-volume sources. With a higher lead price, more effort is put into collecting the scrap lead because more money can be made. There is no mechanism to ensure that scrap lead gets collected to prevent it from entering the environment.

The price of lead is set by international metal markets and is based on the economic laws of supply and demand. For example, as demand for lead increases, if supplies are insufficient to meet the demand, prices go up. As demand decreases, if supplies do not also decrease, prices go down. Since about 1980 the price of lead has been dropping and, in 1986, is quite low. Demand for lead has been decreasing for many years because of environmental control measures, such as reduced amounts of lead in paint, gasoline, plumbing, and printing inks. Demand has also decreased because of the development of alternative technologies that replace lead in certain products with cheaper or more versatile materials, such as copper, plastics, and aluminum. As a result of these reductions in demand, supplies of lead have increased considerably. In addition, further increases in lead supplies have developed because lead ores and production are associated with other metals for which demand is high, notably zinc and silver.

Over the last twenty years, the costs of producing lead and lead-bearing products have increased because of improved pollution control requirements. As a result of the current difficult position of the lead industry (i.e., large supply, low demand, increased production costs), producers and manufacturers tend to consider further controls on point sources or further phase-down (or removal) of the lead content of gasoline as inopportune and inappropriate.

Problems that might arise with further environmental controls include the possibility of lost jobs in some parts of the country, loss of industry revenues and export markets, and additional costs to an already highly controlled industry. The effect of these changes is unequal in different parts of the country. For example, further reductions in, or a ban on, the lead content of gasoline will cause a loss of about 200 jobs[2] in the lead additive industry located in Ontario. However, jobs will be gained in Alberta and Quebec in the refining industry. These regional employment differences linked to the phase-down and phase-out of lead in gasoline will probably show up as a positive influence in Alberta because of the greater concentration in Alberta of both the refining industries and methanol production facilities.

Hidden Costs

There are other costs of lead that are not reflected in the price of lead or of lead products. These other costs are borne by society in many different ways. For example, children with blood-lead levels over 25 μg/dl will have greater health care expenses than those with lower blood-lead levels. Children with lead-induced learning disabilities will also require remedial or special education that is another added expense to society. For some children, remedial education facilities may not even be available to them or may be under-funded or under-staffed. When the most severely affected children reach adulthood, rather than becoming employable taxpayers, they may need some form of social assistance.

Society also pays for lead through government tax dollars that are required to:

- monitor, control, and enforce allowable lead levels;
- research the societal and environmental costs of lead;
- remove contaminated soil from old industrial sites and from some residential yards and playgrounds; and
- educate the public on personal risk reduction techniques and environmental hazards.

Some studies into the health effects and societal and environmental costs of lead are conducted by environmental or other groups acting in the public interest. These groups are also supported indirectly by the public. Costs to the consumer who uses leaded gasoline are higher also due to higher automobile maintenance costs and reduced fuel economy.

All of the costs discussed above are quantifiable. That is, they can (with more or less difficulty) be assigned a dollar value. There are also other large "costs" such as human suffering, lost human potential, and a degraded environment that are impossible to quantify.

All of these "hidden" costs of lead are the economic "externalities" that are not included in the price-determined supply and demand picture. Usually the market price of lead is determined solely by the factors of supply and demand while society ends up paying the true "price" of lead. The lead industry carries on, within a relatively narrow economic orientation, acting as though all economic factors are being taken into account. In these times of economic reappraisal and adjustment, this perspective is clearly distorted and neither the industry nor the national economy is well served.

Governmental intervention in an industry's affairs in the form of costly environmental control measures tends to be interpreted by industries as the addition of "externalities," rather than standard operating expenses.

Anything that increases an industry's operating expenses affects its competitive position with other industries. So long as these control expenses affect all industries equally, there should be no problem. However, this type of equality is very difficult to achieve because no two industries and no two countries are alike. Factors such as location, age of equipment, and existing controls, will differ. As a result, greater control over emissions will require different amounts of investment under different circumstances.

Other factors influence the degree of control on an industry as well. For example, intense public pressure may increase the amount of action taken by governments to enforce legislation and regulations on a particular industrial facility. In some cases, without such public pressure, governmental agencies may not act as swiftly or effectively as necessary for adequate control.

Another factor influencing the quality of lead control by a particular industry is the attitude of management. For example, if the management of a lead-emitting industry accepts its responsibility for maintaining local environmental quality and the necessity for investing in quality emission control, then such control measures can proceed in a relatively straightforward and cooperative manner. If, however, management decides to deny responsibility for local environmental contamination and wants undeniable proof of its responsibility for the contamination before implementing further control measures, then problems arise. These problems can range from poor health for people living near the facility to enforcement difficulties for regulatory agencies. Issues surrounding the enforcement of environmental legislation and regulations are discussed in the following three chapters. Differences in management attitudes toward environmental concerns and control measures can give an unfair advantage to a poorly-controlled industry.

A similar kind of unfair advantage can also be seen in the international marketplace. Environmental control measures and laws are not equally stringent in different countries. Members of the lead industry located in some well-controlled countries are in competition with those in other countries that do not have to deal with the "additional" expenses of high-quality environmental control measures.

A Balanced Perspective

Different perspectives on the costs of lead will yield different conclusions about the wisdom of its continued use. A balanced perspective should consider all relevant factors. The industrial perspective is based on a fairly narrow interpretation of supply and demand. Industries are trapped to a certain extent into a short-sighted outlook based almost entirely on profit and loss balance sheets. Intervention by government in an industry's affairs, in the form of control measures to protect the environment, can seem unfair by cutting into profit margins and/or by putting the industry at a competitive disadvantage.

There are costs to the industry to achieve the desired level of control and there are costs to society if control is not achieved. The main difference between these two types of costs is that the industry's costs are obvious, quantifiable and specific to individual industries, while society's costs are much less obvious, more difficult to quantify, and are borne by the public at large. In addition, industries are far more effective and influential in voicing their opinions or objections about these control measures and associated costs than are consumers and taxpayers.

The resources and action-demanding influences that can be brought to bear on the government by environmental and other public interest groups are rarely as sophisticated, well-funded or effective as are industrial lobbying efforts. In addition, the quantifiable nature of industry's costs allows for a persuasive argument that is easier to grasp and identify with than are the less quantifiable societal and environmental cost arguments. For example, an industry's objections to further environmental controls on point sources or further phase-down of lead in gasoline can be argued in terms of losses in jobs, revenues, export markets, and taxes. In contrast, it requires a much more sophisticated and complex argument to explain the costs to society of health care and remedial education for a certain proportion of children who can be expected to develop learning disabilities if lead in gasoline is not phased out or if a lead industry is not controlled more effectively.

When all of the many cost factors associated with lead are considered, it is clear that hard choices must be made. These choices inevitably involve basic decisions about whether society should continue to pay for the costs and environmental effects of using toxic substances such as lead, or whether the true costs should be borne by the industry that reaps the majority of the financial benefits from producing and using the substance. These hard choices also involve making value judgements. Should we change some or all of our uses of lead to prevent any health effects from occurring? What is a "safe" or "acceptable" level of lead in a person's blood? Questions such as these ultimately become political decisions based on value judgements about what is a fair, responsible and moral approach to a common problem in modern society.

Chapter 12

Governmental Control of Lead

Governmental intervention in the affairs of industries involved in the production and use of lead is essential in order to protect public health, employee health, and the environment. This intervention is essential because lead industries have tended, especially in the past, to treat environmental control and occupational health and safety measures as external to the production process and, therefore, not part of operating budgets. In the absence of governmental intervention, control measures are usually implemented voluntarily only if they also happen to save an industry money.

The impact on human health and the environment from industrial toxins such as lead, creates a conflict between private and public rights. The private rights of an industry involved in the production of lead or lead-bearing products conflict with the public rights of workers in that industry, people living in a nearby, lead-contaminated environment, or the general public at risk from overall high levels of lead in the environment.

In dealing with such conflicts, political decisions must be made, for example, that balance scientific information on health effects against economic information on employment and other industrial cost factors. Government has a responsibility to ensure the protection of public health and the environment while minimizing the possibility of lost jobs or economic damages to a large and useful industry.

Who Controls What?

Since lead is found literally everywhere, control has become an enormous, complicated, and expensive task. Many different departments in all levels of government are involved in monitoring, assessment, enforcement, health effects research, emission control technology research, and regulation development. Local departments of public health may conduct blood-lead testing and public works departments may schedule extra street cleaning in areas around lead-emitting industries.

The federal government in Canada has set standards for the lead content of consumer products across the country. These products include food, leaded gasoline, paints, some electric kettles and glazed ceramic objects, and painted objects intended for children (e.g., toys, pencils and furniture). In addition, the federal government has set guidelines for the amount of lead in air and drinking water, and enforceable standards for emissions from secondary lead smelters and liquid discharges to water bodies from metal mining operations.

In addition to the federal controls, some provinces have standards and guidelines for lead in air, workplace air, dustfall, vegetation, soil, sewage discharges to waterways and sewage sludge applications to agricultural lands. Municipal governments usually control discharges of lead to the sewer system and set "allowable" blood-lead levels. Most of the provinces have adopted the lead standards and guidelines that were first established in Ontario in the early 1970s in response to the controversy over inadequate controls at two secondary lead smelters in Toronto. That controversy and regulatory actions taken around the world at that time, generated modifications and expansions in lead standards and guidelines at all three levels of government. A summary of federal and provincial guidelines and standards for lead in Canada as of 1986 can be found in Appendix J.

Enforceable and Unenforceable Controls

Although the various levels of government have set control limits on lead for a large number of areas, all the limits are not legally enforceable. In general, the terms that are used provide some indication of their enforceability. "Standards" and "regulations" are legally enforceable limits contained in regulations issued under pieces of legislation such as a provincial *Environmental Protection Act* or the federal *Clean Air Act*.

The term "guideline" is most often used when referring to a recommended governmental limit that is not enforceable. Other terms used to refer to unenforceable limits include criteria, objectives, acceptable levels, desirable levels, excessive levels, allowable levels, and intervention levels. Some of these levels can in fact be made enforceable if they are written into specific permits issued by environmental agencies that can be made legally binding on individual industries.

Some unenforceable limits refer only to "action" levels such as the 20 μg/dl intervention (or allowable) level used by the Toronto Department of Public Health for a child's blood-lead level. This level indicates that further investigation is necessary. It would be meaningless to say that it is against the law for a child to have a particular blood-lead level. This interpretation is obviously not the intention of such action levels.

The number of different terms that are used by governments to refer to control limits on lead in the environment can be confusing. It is important, however, to determine whether a level is enforceable or not. If an enforceable level is exceeded, remedial action can usually be taken. Little can be done if the amount of lead is higher than an unenforceable guideline.

Two Approaches to Controlling Lead

Present governmental control strategies are based on two general approaches to pollution control: "control-at-source" and "control-in-the-environment." The control-at-source approach limits the amount of lead released from a specific source, such as a discharge stack, sewer outfall, or sets the amount of lead allowed in consumer products. The control-in-the-environment approach sets maximum levels of lead in air or water with the intention of ensuring an acceptable level of environmental quality. Pollution levels in the environment surrounding an industry are measured and compared to these environmental limits. The adequacy of control measures on this industry can then be judged in light of the measured environmental levels.

These two approaches tend to differ in their enforceability. The source approach generally involves legally enforceable limits that apply to specific point sources or consumer products. The environmental approach sets limits that are "objectives" or "guidelines" for air or water quality and are rarely legally enforceable. When these objectives are incorporated in legally binding permits for individual industries they can take into account site-specific variables such as the number of industries or other sources in an area and the ability of the environment to absorb the pollutant discharges. The development of such permits can be a long and costly process of negotiations among the government, the industry, and, increasingly, other interested parties, including neighbouring residents. In the meantime, unless such legal permits have been developed, unenforceable "objectives" are not very powerful instruments for reducing pollutants.

One exception to this lack of enforceability of environmental limits is the point of impingement standard for air-lead levels (and other pollutants) used in Ontario and Newfoundland. The point of impingement is a mathematically calculated "point" where a pollutant maximally "impinges" (hits) the environment. These standards are calculated in relation to known pollution sources, such as industrial discharge stacks, and are intended to take into account other local variables such as wind speed, wind direction, and local land use that would affect dispersal of stack emissions. Although the standard is enforceable, there are problems that reduce its effectiveness.

These include the accuracy and relevancy of measurement related to a particular industry, complications arising from multiple lead sources in the environment and the fact that it is a control only on the concentration of lead in the air and fails to consider the total loading of lead in the environment.

The many different limits on lead from point sources, in consumer products, and in the environment, may give the impression that lead in the environment is very carefully and closely controlled. The federal government tends to use a control-at-source approach by placing enforceable standards on lead in consumer products and stack emissions from secondary lead smelters. The provinces tend to use a control-in-the-environment approach by setting unenforceable guidelines intended to ensure environmental quality around point sources while using negotiations and legal permits to make excessively polluting industries reduce their emissions. Unfortunately, all these efforts have not resulted in an acceptable level of lead control. Many of the existing limits on lead and the approaches used to establish these controls suffer from some serious problems. These problems are discussed more fully in the next two chapters.

Chapter 13

Problems with Present Controls

Old Limits, Old Information

Many governmental limits on lead in the environment and in consumer products in Canada are based on information that is out of date and consequently, often incorrect. Most of these limits were developed before the mid-1970s. Since then, an enormous amount of research on the health effects of lead has shown that levels considered safe in the early 1970s are in fact "excessive" and an indication of health problems.

The following sections deal first with topics under federal control (lead in gasoline, food, and paint) and then with those under provincial control (occupational exposures, lead in air and dustfall, and soil removal). All of these levels are summarized in Appendix J.

Lead in Gasoline

The most important regulation on lead affecting Canadians is the leaded gasoline standard. Canada's standard for lead in gasoline is the most lenient in the industrialized world. It was set at 0.77 grams per litre (0.77 g/L) in 1976[1] and was relatively high even then compared with other countries with gasoline lead regulations. The Canadian standard will be reduced to 0.29 g/L by January 1, 1987.[2] The U.S. went to the 0.29 g/L level in 1982 and, as of January 1986, lowered it by a further 90% to 0.026 g/L.[3]

In March, 1986, the federal Minister of the Environment, Tom McMillan, announced his intention to reduce the Canadian standard to match the U.S. level of 0.026 g/L by December of 1992.[4] In so doing, Canada will have in place for seven more years, a gasoline lead standard that the U.S. has found unacceptable for children's health. Since the average blood-lead levels in children in the two countries in the mid-1980s are approximately the same (10 to 12 μg/dl),[5] Canada appears to be less concerned than the U.S. about protecting the health of young children.

An average blood-lead level in the child population of 10 μg/dl is described by government agencies in Canada as "relatively low"[6] while in the U.S. such a level forms part of the basis for justifying the need to "reduce the amount of lead in gasoline as quickly as possible."[7] The difference in concern between the two countries over prevailing blood lead levels is striking.

In the U.S., the costs to society of health effects to children with blood-lead levels above 25 μg/dl were quantified[8] (along with other lead-related costs) and found to justify a swift phase-down in the lead content of gasoline. In Canada, the federal government's study on lead in gasoline phase-down options stated simply that:

> "It is not possible to assign a monetary value to human health and therefore a benefit-cost analysis could not be undertaken."[9]

In addition, the Royal Society of Canada's Commission on Lead in the Environment concluded that:

> The reductions in blood-lead levels achieved by these 1987 regulations (0.29 g/L) will, in the opinion of the Commissioners, be sufficient to protect *almost* all segments of the Canadian population against the known harmful effects of lead exposure. There may be exceptions in certain urban hotspots, and among the industrially-exposed labour force.[10] (emphasis added)

No evidence was provided by the Commission to justify this statement. Although the Commission was aware of the fact that 4.2% of children in Ontario had blood-lead levels above 20 μg/dl in 1985, (excluding "hotspot" or urban lead industry dwellers),[11] the Commission did not provide any evidence to show that this percentage would decrease or be eliminated as a result of the 1987 regulation. This statement also gives the impression that children in hotspots need not be included when decisions are being made on levels of lead in gasoline. Even for children in hotspot areas, lead in gasoline is still a major contributor to blood-lead levels. And, *all* Canadian children should be considered when decisions are being made about the wisdom or effectiveness of regulatory changes.

To achieve the limit of 0.026 g/L in the U.S., refineries have either installed additional equipment or increased operating expenses on existing equipment in order to produce gasoline with a sufficient octane level at the lower lead standard. Part of the reason for the delay until 1992 in Canada apparently is to enable the Canadian refining industry to produce gasoline with adequate octane levels without having to rely on an increased use of aromatics, including benzene.[12] However, the future use of aromatics primarily depends upon factors other than the phase-down time frame given to the industry.[13] The most relevant factors are the site-specific nature of inputs, outputs, and in-place equipment at individual refineries. As a result of these diverse factors, the aromatics content of gasoline will not likely be affected by a seven year (Canadian) rather than a two year (U.S.) time frame. What would most clearly affect the amount of aromatics in gasoline would be specific regulations limiting their amount.

The Canadian government is taking a curious approach to the leaded gasoline problem. Rather than establish a tight phase-down time frame or restrictions on aromatics content, they are conducting an enormous public education campaign focussing on misfuelling. The fact that the refining industry is sharing in the funding for this program might help to explain why such a campaign is occurring.

Although everyone is in agreement that misfuelling is unwise because it increases several kinds of pollution, public education on misfuelling ought to be used as a *supplement* to more substantive and swift regulatory changes.

One type of change that would probably stop misfuelling overnight is to equalize the at-pump prices for leaded and unleaded gasoline. The price differential is three to four times higher than the actual manufacturing cost differential.[14] Studies in the U.S. show that misfuelling will not be significantly reduced until the price of leaded fuel is as high (or higher) than that of unleaded fuel.[15] The dollar-for-dollar contribution[16] by the refining industry to the public education campaign on misfuelling appears impressive. However, this amount of money is far less than the industry would lose if they discontinued the practice of falsely elevating unleaded gasoline prices.

Equalization of the at-pump price of leaded and unleaded fuels can be handled at the provincial level by implementing changes in tax structure or by legislating equal prices. Representatives of the refining industry are quick to point out that equalization via tax changes probably would be thwarted by a free marketplace resulting in prices reverting to the price differential which has been maintained over the past few years. However, the fact remains that such a price difference is larger than the actual manufacturing cost differences. The most logical approach to equalized prices would be for both industry and government to play a role. Industry could reduce the at-pump price difference to actual production cost differences. Government could adjust the tax structure as necessary.

Lead in Food

The federal regulations for the lead content of food[17] were set in 1968 on the basis of the scientific information then available. In 1968, blood-lead levels above 80 μg/dl in children were considered "dangerous." Research since then has documented very serious health problems at blood-lead levels well below 80 μg/dl. Since the present "elevated" blood-level is 25 μg/dl for children, it seems clear that limits established to prevent blood-lead levels from going over 80 μg/dl may not be at all adequate to keep blood-lead levels below 25 μg/dl.

A further problem with food-lead levels is that the regulated foods are not representative of many of the foods consumed in Canada. For example, the regulations do not mention milk and milk products, meats other than liver, breads and cereals, or canned foods. Canning that involves a lead-soldered seam has been shown to increase food-lead levels by up to ten times.

Table 13.1

Comparison of Legal Limits and Actual Levels of Lead in Food

Food	**Legal Limit** (ppb)	**Actual Level** (ppb)
marine and fresh water animal products	10,000	200-600*
liver	2,000	50-100**
fresh fruit	7,000	20-140
fresh vegetables	2,000	30-70
tea	10,000	30

* Great Lakes Fish
** beef, pork, lamb, poultry

Sources: *Legal limits* ***Food and Drug Act,*** *R.S.C.,1970 B.15001; actual levels Nutrition Foundation Inc. "Assessment of the safety of lead and lead salts in food: a report of the Nutrition Foundation Expert Advisory Committee." Washington D.C. 1982; Great Lakes fish—J. O. Nriagu, "Lead contamination of the Canadian environment," unpublished report to the Royal Society of Canada's Commission on Lead in the Environment, 1985.*

Table 13-1 compares the regulated legal limit and the typical lead content of some foods consumed in Canada. As can be seen in this table, the regulations are many times higher than the amounts of lead that are actually found in food. Since most foods are not covered by the regulations, and since the regulated levels are based on outdated health information, assurances from government inspectors that foods are within the regulations for lead are meaningless. An updating of food-lead regulations is very important since food is one of the principal lead-exposure pathways for most people.

Lead in Paint

The federal regulations for the lead content of paint were implemented in two stages. Leaded paint on toys, furniture and other articles intended for children was regulated in 1970 at a maximum level of 5000 ppm.[18] At that time, a blood-lead level up to 60 μg/dl in a child was not considered a health problem. In 1976, all paints for use on the interior or exterior surfaces of buildings, furniture or household products were required to be at or below 5000 ppm of lead.[19] Only one year later, the U.S. reduced its lead in paint standard to 600 ppm[20] in order to avoid or reduce health effects in young children. Canadian paint lead regulations were adopted and are still based on health information that we now know is inaccurate. As of May, 1986, the Canadian limit is more than eight times the U.S. limit.

Occupational Lead Exposure

For maximum unprotected (without a mask) exposure to lead, several provinces across Canada use an occupational limit of 150 micrograms of lead per cubic metre of air ($\mu g/m^3$) over an eight-hour day, 40 hour week.[21]

The 150 $\mu g/m^3$ level was set in the early 1970s in the United States.[22] This level is intended to avoid lead poisoning in workers and ignores the many known health effects that can occur at lower lead-exposure levels. The occupational air-lead regulation is generally accompanied by "target" blood-lead levels for workers, indicating some degree of "too much" lead. For example, workers are removed from occupational exposure when blood-lead levels reach 60 $\mu g/dl$ (in British Columbia)[23] or 70 $\mu g/dl$ (in Ontario)[24] and cannot return to a lead-exposing job until blood-lead levels are at or below 40 $\mu g/dl$ (in British Columbia) or 50 $\mu g/dl$ (in Ontario).

The World Health Organization (WHO) recommended in 1980 a "health-based biological exposure limit" for blood-lead levels in lead workers of 40 $\mu g/dl$ for males and females over the reproductive age and 30 $\mu g/dl$ for women of child-bearing age.[25] In both Canada and the U.S., acceptable blood-lead levels in occupationally exposed individuals are higher than the WHO recommendation although they are more carefully controlled in the U.S. than in Canada. In the U.S., workers are required to leave the lead-exposing workplace if their blood-lead level reaches 50 $\mu g/dl$ (compared to 60-70 in Canada) and may not return until they are at or below 40 $\mu g/dl$ (compared to 40-50 in Canada).[26]

WHO also recommended that, depending on the pre-employment blood-lead level of employees, limits on air-lead levels ought to range from 30 to 60 $\mu g/m^3$ (values that are much lower than the Canadian 150 $\mu g/m^3$). The occupational lead-exposure situation is considerably different in the U.S. than in Canada, in that the U.S. exposure limit was revised downward to 100 $\mu g/m^3$ in 1981 as an interim standard until June 1986 for secondary lead plants and until June 1991 for primary lead plants. As of those dates, the interim standard would be lowered further to 50 $\mu g/m^3$,[27] which is in line with the WHO recommendation. The current Canadian standard of 150 $\mu g/m^3$ clearly provides less protection for the Canadian lead worker than is provided for his U.S. counterpart.

In January, 1986, the Ontario Ministry of Labour drafted a proposed regulation[28] for construction workers from exposure to lead-paint dust and fumes during sanding, burning, sandblasting, and flame-cutting of materials coated with lead paint. The regulation covers proper respiratory protection at various levels of exposure, care of respirators, education about lead hazards and techniques to avoid exposure, containment and clean-up procedures, warning signs on the work-site, medical care and blood-lead testing.

Although this proposed regulation appears to be comprehensive, there are two problems. One, the blood lead level for excessive exposure is the same high level used in the existing occupational lead regulation. The second problem is the requirement that the various precautions listed above be followed when working on materials that "contain or may contain lead" without any requirement for testing the lead content of the materials. The Ministry of Labour assumes that supervisors will know when lead is present. Under such circumstances, governmental inspectors will have to be extremely attentive and responsible to ensure that construction sites are being managed properly so that workers are protected. This may not be possible since there are typically very few inspectors to cover the many thousands of workplaces under the Ministry's jurisdiction.[29]

Lead in Air

Provincial controls on point sources of lead tend to concentrate on air emissions and, with the exception of British Columbia, follow the levels set by Ontario in the early 1970s. Ontario's criterion for air lead of 5 $\mu g/m^3$ is an air quality objective for a 24-hour period.[30] The point of impingement standard[31] (10 $\mu g/m^3$/half hour) is mathematically derived from the 24-hour criterion and is, therefore, based on the assumptions involved in the setting of that criterion. All of the assumptions[32] that went into the setting of the 5 $\mu g/m^3$ level can now be seen to be either inaccurate or inappropriate, particularly when the health of young urban children (or young children living near an urban, lead-emitting industry) is considered.

Examples of inaccuracies in these air-lead assumptions include incorrect data on: the amount of lead in the diet, the absorption rates for food and air lead in children and for air lead in adults, and the amount of lead that could be safely consumed each day by children and adults.[33] Because of the availability of new information, it is clear that, at a minimum, the air lead criteria should be re-examined and may in fact, need to be lowered.

Lead in Dustfall

Ontario is the only province with a guideline for the level of lead in dustfall around a lead-emitting industry. This level was set in 1974 in response to the problem of heavy lead-dust fallout around the Toronto lead smelters.[34] The unenforceable guideline is 100 milligrams per square metre over a 30-day period (mg/m^2/30 days). The Ministry of the Environment staff who developed this level had very little information to base it on and recognized that as more information became available the level would probably have to be lowered.[35] As of 1986, it has not been changed even though new, relevant information exists (e.g., rapid rates of soil lead accumulation and the major role of dust and soil lead in determining children's blood lead levels).

The dustfall guideline has been regularly exceeded around one Toronto lead plant since the early 1970s, except for a couple of relatively "clean" years in the late 1970s.[36] These exceedances ("exceedance" is commonly used by experts to denote a level in excess of a governmental standard or guideline) have resulted in excessive soil-lead buildup in some nearby residential yards. For example, one assumption on which the guideline was based stated that soil-lead accumulation would be limited to an increase of only 300 ppm over a twenty-year period.[37] In fact, actual soil-lead levels have been shown to increase by more than 300 ppm in *one* year in a backyard next to the smelter.[38] The much faster soil-lead accumulation in this backyard may be due to the problem of regular exceedances rather than with the actual level chosen for the guideline. However, the fact remains that the existence of an unenforceable guideline has not prevented dramatic increases in soil-lead levels.

Lead In Soil

The soil-removal guideline used in Alberta, Manitoba, and Ontario of 2600 ppm is also based on outdated information. The justification for this level comes from one small study on rats done in 1980 on the amount of lead that can be absorbed from soil with 1000 and 3000 ppm of lead.[39] These data showed that at soil-lead levels above 1000 ppm a relationship began to emerge between elevated soil-lead levels and elevated blood-lead levels, but that this relationship was only fully clear at 3000 ppm. Because there is a potential measurement error factor of 13% when soil-lead levels are being determined, a soil test result of 3000 ppm could be anywhere between 2600 and 3400 ppm, and a measured level of 2600 ppm could actually be 3000 ppm. Therefore, the 2600 ppm removal guideline was chosen on the basis of avoiding the known effects of elevated blood-lead levels at 3000 ppm minus the measurement error factor. This method uses an approach that allows for no safety margin and ignores the possibility of risks at lower exposure levels.

More recent research shows that soil or dust-lead levels at much lower levels (i.e., 500 to 1000 ppm) appear to be responsible for increased blood-lead levels in children.[40] This is in fact verified by circumstantial evidence of blood-lead levels in children living within the 1000 ppm soil-lead isopleth around a secondary lead plant in Toronto. The isopleth is used by the Toronto Department of Public Health to delineate the "designated area" within which annual blood-lead testing is done. This particular smelter achieved a high degree of emission control by the early 1980s. Since that time, blood-lead levels have appeared to stabilize at consistently elevated levels pinpointing the soil as the most likely ongoing exposure source.

Another recent study found that soil-lead levels contribute to house dust-lead levels by a factor of 2.2.[41] This finding means that soil-lead levels considered "safe" at 2000 ppm could contribute to a house dust-lead level of 4400 ppm. This level is four to eight times higher than dust-lead levels that are associated with increases in children's blood-lead levels. On the basis of all of this recent information, a soil-removal guideline of 2600 ppm is too high to provide adequate health protection. See Appendix I for recommended regulatory changes.

Fortunately, most regulations and objectives in Canada for lead in food, air, and soil are exceeded only rarely if at all. However, assurances that government regulations or guidelines are not being exceeded tells us very little about the safety of lead exposure. Because of an inaccurate, outdated information base, many of these levels are simply too high. Even actual levels of exposure, as reflected in current average blood-lead levels, provide no real safety margin below the level at which obvious health effects can occur. Lead levels in the environment need to be reduced and the regulatory structure for dealing with them needs to be modernized.

The "Dilution Solution"

The two control approaches discussed in Chapter 12, control-at-source and control-in-the-environment, suffer from a common problem. Neither approach places any overall limit on the amount of lead in the environment or in consumer products. The control-at-source approach would seem to be able to limit total amounts but in fact only limits the *concentration* of pollutants in a discharge stack, sewer outfall, or consumer product. Under such an approach, any amount of lead can be released from a point source to the environment so long as it is sufficiently diluted to meet the concentration requirements. For example, a point source can release a very small amount of lead from its stack, or a very large amount and keep the concentration the same simply by diluting the airstream with a greater amount of air. In the same way, a large industry will release more lead than a smaller one simply because a larger operation can have more discharge stacks than a smaller one. More lead is released from the larger operation because the limits are placed on each stack rather than on the entire operation.

The control-in-the-environment approach does not place any total limit on lead entering the environment either. Rather, this approach allows for widespread disposal of an industry's emissions by using natural wind movement and dispersal. For example, if a point source of lead emissions has a very tall discharge stack, it can release a very large amount of lead because stack emissions released high into the air can be widely dispersed into the environment, thus keeping the concentration at any one point in the local environment below an acceptable level. The taller the stack, the more lead that can be released while still meeting the government's control-in-the-environment limits in the area surrounding the industry.

Both approaches, control-at-source and control-in-the-environment, involve the "dilution solution to pollution." The "dilution solution" assumes that any amount of a contaminant can be released to the environment so long as this release occurs slowly and gives the environment time to absorb the pollutant and spread it out over a wide area at harmlessly low levels. Unfortunately, this "solution" ignores some very important properties of many pollutants, such as their persistence and ability to accumulate in the environment, and their interaction with other environmental factors after they are released (e.g., sulphur dioxide contributes to acid rain formation). Difficulties with this type of "solution" may surface later or in other locations often in a form that is larger and more difficult to deal with. For example, since lead does not break down, levels that are relatively safe when they are released, may accumulate to dangerous levels.

For many years governmental control of lead has concentrated almost exclusively on air-lead levels. This focus is inappropriate because it ignores the accumulation of lead in the environment and the fact that most of the particles in industrial fugitive emissions are too large to show up in air-lead monitors. It also ignores the fact that lead particles in soil or dust can be redistributed into other exposure pathways (e.g., food and house dust).

The most important of all these criticisms is the fact that air-lead measurements do not give a picture of total lead fallout and environmental accumulation.

A more appropriate technique and control strategy than air-lead measurements involves measuring total lead fallout over a longer time period. This kind of measurement, which is now being made in a number of locations, uses open dust fallout collectors to record the total lead content for a week or month. This type of information can reveal the extent of the exposure problem in any location in a way that is more relevant than air-lead data. For example, lead in dustfall at one location in a downtown residential area of Toronto occurs at the rate of about 14,000 micrograms per square metre per 30 days ($\mu g/m^2/30$ days).[42] This rate represents a daily fallout of about 475 μg for each square metre.

If we consider a typical inner city backyard of about five metres by four metres (i.e., 20 square metres), the amount of lead fallout on this yard would be about 9,500 μg per day. A child playing in this yard would not have to do anything very unusual to stir up some lead dust and breathe it in, or put a dusty finger in his or her mouth, or eat lunch without washing every bit of lead dust off his or her hands. Any of these actions could put dozens or hundreds of micrograms of lead inside the child's body.[43] Even if the backyard was only one square metre, the child could be exposed to over 400 micrograms of lead falling into the yard every day. Compare these amounts to the total amount of lead in air that a child would be exposed to playing in this same yard. The average lead content of air in an inner city neighbourhood is about 0.54 $\mu g/m^3$.[44] Since a child inhales about 6 m^3 of air per day, he or she is exposed to about 3.24 μg of lead simply by breathing. It is clear that a daily lead dustfall of 9,500 μg provides a much larger exposure potential for the child, and, therefore, should be a much more important concern to regulators of lead, than the amount of air lead the child inhales.

The rate of lead fallout is probably the most important factor to consider in developing control strategies since it has the greatest relevance for blood-lead levels. Lead fallout contributes directly and indirectly to lead in food and dust, the two most important contributors to blood-lead levels. Concentrating lead control efforts on the primary emission sources (e.g., vehicles using leaded gas, fugitive emissions from lead industries, and old caches of lead dust or lead paint in homes) would limit environmental lead contamination much more swiftly and effectively than has been done so far.

It is important to remember that the current environmental lead burden results from a long history of lead use. Because lead is persistent, current exposure levels are a combination of these historical accumulations and current emissions. Developing improved control strategies now will decrease the addition of new amounts of lead into the environment, but it will be many years before environmental lead levels become so low that no one needs to be concerned about possible health effects. For this reason, the personal risk reduction lifestyle choices described throughout this book are important now and will continue to be for many years.

Chapter 14

Legal Angles

Common Law and Statute Law

Legal issues surrounding lead pollution can involve both of the two main types of law: common law and statute law. Common law has been developed over the years by judges deciding on individual court cases. These decisions become "precedents" for later decisions. Common law protects persons or property from damage and can be the basis for obtaining compensation for injury to health or property damage.

Statute law includes all legislation passed by provincial legislatures and the federal parliament. Judges interpret and apply these statutes in individual court cases. As with common law, decisions made by judges become precedents that will be considered when later decisions are made in similar cases.

The wording of the statutes (legislation) is usually very general. For example, pollution of the environment or workplace by contaminants is "prohibited" and the government is given the authority to pass specific regulations dealing with individual pollutants. When a regulation for a specific substance has been fully developed, approved, and promulgated (publicly announced), it includes maximum allowable pollutant levels usually called standards, and often specific rules about industrial operations, such as how often the pollutant level must be measured. In addition to regulations passed under specific statutes, environmental or occupational safety and health agencies often develop "guideline" levels to control pollutants. These levels are set internally by these agencies, do not require governmental or cabinet approval, and have no legal force.

Three legal requirements are necessary to file a lawsuit in an environmental matter. The defendant (the polluter) must have allegedly committed some violation of common law (e.g., created a nuisance or an act of negligence) or statute law (e.g.,violated the general prohibitions in an environmental statute or exceeded a regulation). Second, the court hearing the case must have jurisdiction (or authority to act) over the parties involved and the offense in question. Third, the plaintiff must have "standing," which involves having a special interest in the matter beyond what members of the general public would have, such as a violation of personal or property rights.

After the above three requirements are met, there are two kinds of proof requirements in law, depending upon whether the case is a criminal or a civil one. In criminal and quasi-criminal cases (where a statute, a regulation, or a legal permit issued under a statute has been violated), the accused is innocent until proven guilty beyond a reasonable doubt. In civil cases (where the common law has been violated), the plaintiff charges the defendant with *liability* instead of guilt. Liability (i.e., being legally responsible for damages) need only be established on the basis of a "balance of probabilities" rather than on proof beyond a reasonable doubt. This means that the plaintiff must give evidence to indicate that there is a reasonable probability that the accused is responsible for the action or effect under consideration.

The Problem of Evidence

When an environmental regulation is breached, locally affected people, or, more appropriately, the government acting on their behalf, should prosecute. Proof is required to show that the accused was responsible. The kind of proof or evidence needed depends on how the regulation is worded. Regulations containing maximum limits established under the control-at-source approach are easier to enforce than those under the control-in-the-environment approach because the source approach is based on actual measurements. With the source approach, it is a simple matter of periodically measuring the lead content in the discharge stack or product. If the limits are exceeded, the company is prosecuted, or goods are seized and manufacturers, distributors, or importers are prosecuted.

Maximum limits established under the control-in-the-environment approach are usually unenforceable guidelines. Their primary use by environmental agencies appears to be in negotiations with an industry to achieve effective emission control. However, some environmental limits are enforceable standards, but tend to have limited legal enforceability (e.g., Newfoundland's and Ontario's point of impingement standards).

Limitations in the enforceability of control-in-the-environment standards stem from problems of measurement accuracy and relevancy. For example, the standard is a maximum lead concentration at the point of impingement. However, the concentration of air lead is not routinely measured right at the point of impingement. Instead, air-lead concentrations are measured by monitors at various locations around a lead-emitting industry. The location of air-lead monitors is not determined by the point of impingement location, or by any specific guidelines or standards, but is usually a compromise between ideal locations and what is actually available.[1] Since measurements are not routinely taken at the point of impingement, air-lead concentrations obtained from monitors do not relate directly to the enforceable standard.

Point of impingement standards are not used on a routine basis to ensure compliance by an industry during its regular operations. Instead, the standards are used primarily during the issuance of Certificates of Approval. These legally binding certificates are required when a company begins operations or changes existing operations. Under these circumstances, there may be no prior measurements to show how much contamination could arise. Mathematical models are used to calculate the emission levels that will satisfy the point of impingement standard. The models are intended to take into account as accurately as possible local variables such as stack height and weather conditions that would influence dispersal.

A further problem that makes the violation of legally binding environmental limits more difficult to use as evidence than the violation of source limits is the fact that measurements around a lead-emitting industry can include lead from other sources, such as other industrial emissions, automobile exhausts, or lead particles from contaminated soil blown by the wind. Because of these multiple possible sources, it becomes difficult to successfully prove a violation based only on air-lead measurements. Usually further evidence is required, such as records of equipment malfunction or an eyewitness account of an unusual dust cloud coming from the stack. There are however, new modelling techniques being developed that can apparently distinguish multiple sources and more effectively isolate individual source contributions.[2]

Those environmental limits that have not been developed into enforceable standards sometimes can be made legally binding by incorporation into, for example, discharge permits (British Columbia), orders (Manitoba), or control orders (Ontario) on individual companies. Permits such as these for individual companies are developed to account for site-specific variables, such as the number of industries in an area and the ability of the environment to assimilate multiple pollution sources. Technical and economic factors are also considered. Development of permits with strong environmental control requirements is relatively rare primarily because of a company's right to appeal the conditions of the permit in the courts.

By the time an environmental regulation is violated and enough evidence has been gathered to successfully prosecute, environmental degradation and/or human health problems already may have occurred. The problem is that environmental laws and regulations tend to react to pollution problems after they arise rather than preventing their occurrence. This same reactive approach can be seen in the "effects based" approach used in the setting of pollution control limits.

"Effects-based" Control Limits

In general, the setting of enforceable control limits requires proof that exposure above these limits would cause harm. Without this proof, these limits would be open to legal challenge. One of the reasons that guidelines are unenforceable is because they are usually based on information that does not provide extensive enough evidence of proven harm. Just as people are "innocent until proven guilty," pollutants tend to be given this legal right as well, because in court they are associated with the actions of an individual or corporation.

Several problems arise when pollutants are considered "innocent-until-proven-guilty." The only health risks that can be considered using this approach are those that cause an observable health problem. It is an "effects-based" approach, which means that only known, proven human health or environmental effects are considered. Although safety margins are sometimes included in governmental control limits, these limits are usually only unenforceable guidelines or suggested levels for safe exposure.

The assumption that all effects of a pollutant are known and can be quantified ignores the seriousness of unproven or difficult to prove effects at lower exposure levels (i.e., below threshold effects) that may be able to cause harm. For example, in the case of lead, there may be changes in a child's I.Q. at blood-lead levels below which there is any indication of obvious adverse health effects. Damages can occur while scientists are attempting to "prove" their occurrence. Lead-induced changes in a child's behaviour and brain waves may be occurring and may even be irreversible at levels below those where other "proven" health effects occur.

The persistency of lead in the environment is also ignored by the "effects-based" approach. Control limits are based on proof that a contaminant will cause harm at the concentration level at which it is released (control-at-source) or allowed in the environment (control-in-the-environment). Exposure levels that may be safe when the lead is released may become dangerous as the lead accumulates in the environment.

One of the biggest problems with an approach that assumes that pollutants are "innocent-until-proven-guilty" is the inability to prevent damage. If control limits are exceeded and a polluter is successfully prosecuted, health problems or environmental damages probably have already occurred. This problem is offset somewhat by more recent attempts at preventive control. For example, environmental agencies generally have in place an approvals process that is intended to prevent excessive environmental contamination by requiring prior approval from the regulatory agency. However, since this process only applies to new or changed operations, there are still many polluting companies in existence that are not as appropriately controlled.

A further factor ignored by a reactive, rather than preventive, approach is the difference in economic costs between preventing and cleaning up a problem. Clean-up can be extremely costly, and sometimes impossible, after environmental degradation has occurred. These costs are usually much higher than the costs of preventing the emissions in the first place. Prevention not only reduces the costs of environmental clean-up, but it also avoids the development of damages that may be irreversible.

Another approach to setting control limits is based on the "best available technology" for pollution control. For example, the control limits on secondary lead smelters[3] set by the federal government under the *Clean Air Act* are based on the use of the best available technology. However, health effects can play some part even with this approach since the need to set the limits in the first place often results from evidence of known health effects in the areas around such smelters. Economic factors (i.e., economic feasibility) also can play a role in the setting of regulations or other control limits. However, the control technology that is economically feasible may not necessarily be the best control technology available.

Modernizing the System

Governments have a credibility problem on lead issues because many of the limits they have set:

- are based on out-of-date and consequently often incorrect information;
- do not provide any safety margin;
- are worded in a manner that is difficult to enforce;
- are, in some cases, not enforceable at all; and
- carry maximum fines that are generally too small to serve as a deterrent.

In addition, neither past nor current standard-setting procedures provide much room for public involvement.

The enforcement and regulatory efforts of environmental agencies will not be able to meet public expectations if an outdated information base is used. Research on environmental and health pollutants, such as lead, is a dynamic field. The most recent research represents a kind of scientific "frontier" constantly building on accumulated evidence from the past.

A simple way to deal with the reality of this ever-changing and ever-growing information base is to establish a mandatory review process that would place a legal obligation on regulatory agencies. Under such a mandatory review process all standards and guidelines could be re-evaluated regularly in light of new information and brought up to date where necessary. Standards could also be developed on a regular basis for newly recognized pollutants of concern.

Such a process now exists in the U.S. Environmental Protection Agency for all air quality standards, which must be reviewed every five years. In these reviews, the most recent scientific evidence is used not only to determine adverse health effect levels, but also to incorporate appropriate safety margins to account for the possibility of risks at lower levels. If this kind of mandatory review were established in Canada for lead regulations, we would probably find that existing standards, guidelines and control approaches are out of date and overly simplistic.

Legislation, regulations, standard-setting, and the enforcement policy of environmental agencies are not written in stone. The rules must be flexible enough to change when circumstances change or when new information develops.

The Ontario Ministry of the Environment is addressing some of these problems in its current review of the provincial general regulation for air pollutants.[4] This review is an important step towards needed regulatory reform. It was instituted in recognition of the fact that this agency needed to increase its credibility and enforcement capabilities.[5] The Ministry has pointed out in the early stages of this review that in order to get away from the dilution solution and its associated problems, it must establish a policy of controlling pollutants at their source and setting control limits that account for total *loading* of pollutants in the environment. An important point raised in the preliminary stages of the process is the need to classify pollutants on the basis of important, health-relevant characteristics. For example, control strategies should be different for substances that can cause cancer, that are persistent, or for which limited information exists.

The persistency of lead is one of the most important characteristics that should be kept in mind when lead regulations are modified or developed. As already discussed, persistence makes the use of both the "dilution solution" and an immediate "effects-based" approach inappropriate. A further factor related to lead's persistency is the selection of the most appropriate media to monitor and control. The present emphasis on air-lead levels does not deal adequately with the fallout of lead onto soil, dust, and food.

A further problem with environmental legislation in Canada is that federal and provincial Ministers of the Environment and their senior officials are not charged with a legal duty to enforce the law. They *may* act or are only required to act *at their discretion*. Inclusion of a legal duty to enforce the law would strengthen the operation of the regulatory control system and could be included in the legislative changes accompanying the setting of a mandatory review process.

Members of the public, either as individuals or as organizations operating in the public interest, are increasingly demanding more involvement in governmental control of pollution. These demands are appropriate and greater public participation is valuable. The risks and scientific uncertainty associated with certain contaminants mean that decisions on allowable levels are more than just scientific or administrative matters. They are also based on value judgements about what is tolerable or acceptable and as such, they are political decisions.[6] Political decision-making is most effective when all people affected by the decision, and interested in participating, have full and equal access to all relevant information and an early opportunity to participate.

Greater public participation in decision-making represents an important change in attitude towards making elected officials and their staff more accountable to the public. The public has an important watchdog role to play. For example, the above recommendation for a regular, mandatory process of reviewing existing standards and establishing new ones could produce weaker standards if this process is not open to public scrutiny and peer review.

Table 14.1

Modernization Steps

1. Bring the regulatory structure up to date by required periodic, mandatory reviews of standards and guidelines.
2. Charge senior environmental officials with the legal duty to act in accordance with the terms of legislation and regulations.
3. Replace the "dilution solution" principle with a control-at-source approach that accounts for total loading of persistent pollutants.
4. Adopt a preventive approach (including safety margins for health issues) by setting control limits that avoid damages rather than applying control after damages have occurred.
5. Make as many as possible of the unenforceable controls legally enforceable.
6. Improve the enforceability of regulations by using monitoring methods that are directly relevant to the terms of the regulation.
7. Develop stronger enforcement penalties, including realistic fines and imprisonment.
8. Demand proof of safety from industry rather than proof of harm from the public.
9. Acknowledge the role of value judgments in decision-making by permitting fuller and earlier public participation in the development of control strategies.

Finally, in addition to a review of governmental control limits and strategies, there is an important legal issue under review and in need of modernization. This is the issue of the "burden of proof" (or onus of proof). It can be prohibitively expensive for private citizens or environmental groups to obtain necessary legal and scientific proof in environmental matters. Considerable discussion has occurred about where the burden of proof should lie in these situations. For example, questions arise about whether a citizen (or group) should have to prove that a substance or action is harmful or whether the manufacturer (or user) should have to prove that no harm is involved. In criminal and quasi-criminal cases, the polluter is innocent until proven guilty. However, in civil cases where liability must be shown on the basis of evidence relating to a balance of probabilities, environmental lawyers have recommended a reversal of the burden of proof. Such a reversal would require the manufacturer or polluter to prove that the product or action in question will not harm human health or the environment. For a fuller look at this kind of legal reform, see the "Environmental Bill of Rights" in Appendix E.

Political and Economic Obstacles

There is always a great deal of inertia that must be dealt with when major changes, such as those described above, are under consideration. There is inertia in the general public, in the bureaucratic structure, and, perhaps most importantly, in those industries who will be affected economically by the changes. This latter group usually presents the greatest obstacle to the implementation of changes in legislation and regulations. Ultimately, changes of this sort are subject to careful political scrutiny and generally require the combined approval of members of federal or provincial cabinets. Access to this level of government is easiest for those who have developed a strong lobbying system.

Industrial lobbying efforts can significantly influence the ability of environmental agencies to implement regulatory reforms. A good example is the ten-year debate over the phase-down of lead in gasoline. This debate began as a scientific controversy. Expert opinions were presented with equal intensity on opposite sides of the question of the effects of leaded gasoline emissions on children's health. This scientific controversy has now been resolved. Leaded gasoline is the primary contributor to environmental lead contamination and blood-lead levels. The current debate is over the time schedule of the phase-down.

The cost-benefit analysis of lead in gasoline conducted by the U.S. Environmental Protection Agency was used to justify a swift phase-down schedule in the U.S. simply in order to save sizeable amounts of money for the nation as a whole while considering the costs to the industry to be reasonable and affordable. The phase-down time schedule was not overly abrupt since it allowed the industry a reasonable amount of time (two years) to develop the necessary modifications and operational re-tooling. It is reasonable and fair to allow this kind of time when phase-down schedules are developed, but it is not fair or reasonable for a comparable industry, i.e., Canadian refineries, to be given six or more years for what could be accomplished in a much shorter amount of time.

In Canada, no such cost-benefit analysis has been done. The federal government does not seem to be considering all sides of the economic picture, apparently convinced that the industrial cost penalty of a swift phase-down is unattainable and unfair to the industry. However, the cost penalty to individuals and to public funds has not been fully considered. The debate, therefore, continues over how to implement obviously necessary regulatory reforms that happen to conflict with powerful vested interests.

PART 3 ACTION

Chapter 15

How to Begin Making Changes

Risk Reduction for You and Your Family

Now that you are aware that lead could be increasing your personal risk of health problems, the first thing to do is to evaluate your situation for the degree of such risk. People in the following categories should take personal responsibility for reducing their degree of exposure to and absorption of lead:

- women who are pregnant or who ever plan to have children;
- people who are exposed to lead at their workplaces;
- parents, guardians, teachers, and all people who are responsible for children under the age of six years;
- people who live next to or near, a heavily-travelled street or roadway;
- people who live in an area where there may be high environmental lead levels in the dust, soil, or air because of the presence of an industry that uses lead; and
- people who live in the inner core of a large city.

The first step in reducing your risk is to find out just how large your risk might be. Contact your provincial ministry of the environment (or other governmental agency as appropriate) for information on lead levels in the air, soil, dust, and water in your area. Check Figure 29 in Chapter 9 for the estimated blood-lead level for someone in your category. If you are one of the persons in the above categories, you might want to request a soil test for the amount of lead in your garden and lawn. Consider your usual diet to see if it is high in canned, acidic foods. If, on the basis of the information you have, you think there is a possibility that your blood-lead level (or that of your child) might be a little high, ask your local doctor or health centre for a blood-lead test. In most places, a blood test is covered by regular medical insurance programs.

The second step for those who decide that their risk level needs to be lowered is BEGIN REDUCING YOUR RISKS. Look back at Chapter 3 and choose those risk-reduction actions that are appropriate for your situation. For example, a very important and quite simple action is careful and thorough washing of your hands, or those of a child, every time before you handle or eat food.

Set a good example and train children to follow this example. Some points to remember include: never eat anything that falls onto the floor or ground; wash or peel food before eating or cooking; wet dust and wet mop your home or childcare centre frequently; and eat foods high in calcium and iron.

You may be the only person in your family or among your friends who has become educated on lead risks. Share your information. Have a family discussion. Explain the serious and long-lasting effects that lead can have on the quality of life and health, especially for young children. Describe how you get exposed to lead and what you can do to reduce your exposure and absorption. Figure out some practical changes for each member of the family and emphasize the importance of their personal responsibility for reducing the lead risk for themselves and for others. Make your lead risk reduction changes fun and important. Get in the habit of reminding each other frequently of your new lifestyle habits and call attention to the progress you are making. Be on the lookout for any special events, such as cleaning out a dusty attic or undertaking some renovations, that could temporarily raise lead levels considerably.

If anyone in your family has a blood-lead level that you want to lower, practice your new lifestyle habits for about three months and then get their blood checked again. It should be lower and will slowly keep on getting lower as you continue your new habits.

Suppose the water tap in your kitchen was turned on enough to cause a pool of water to form on the floor. Mopping the floor would clean up the water for a short time, but unless you turned off the tap, the floor would get flooded again. However, once the tap was turned off, mopping up would return everything to normal.

Environmental contamination is similar. First the sources must be "turned off" (that is, emissions of contaminants to the environment must be stopped or reduced to an acceptable trickle) and then contamination in the environment can be cleaned up (for example, lead-contaminated soil in residential yards can be removed and replaced with clean soil).

For lead, there are two primary "taps" to be turned off: industries and automobiles. How to turn these off is discussed in the next two sections. Public participation in these processes is discussed in the final section in this chapter.

"Turning Off the Industrial Tap"

The first step in reducing industrial emissions is to identify the emission problems for a particular industrial facility. The type and extent of lead contamination around that industry provides most of the necessary clues. Since lead tends to accumulate in soil, soil-lead levels have acted as a kind of "thermometer" to indicate the location and extent of contamination around an industry. If the lead contamination extends for a considerable distance from the plant and is not particularly severe very close to the plant, then smoke-stack emissions are the likely problem. On the other hand, if the lead contamination problem is primarily quite close to the plant then the problem is more likely to be fugitive emissions.

The primary control procedure for excessive stack emissions is a baghouse, which traps the lead dust in filter "bags" before it enters the stack. This kind of emission control technology is found in lead industries across Canada. Some industries also have a back-up baghouse for use in the event of failure of the primary baghouse.

Improving the quality of stack emissions depends on three factors: the existence of an appropriately sized and well-maintained baghouse system; a back-up baghouse that can be used if there is a breakdown or failure in the primary baghouses; and well-trained, well-motivated, careful employees. In order for a baghouse to remove lead particles effectively, it must be kept in good condition and the exhaust fumes entering it must not be too moist. Since the degree of moisture in the exhaust fumes is determined primarily by furnace operating procedures; human experience, judgement, and motivation play important roles in the effective functioning of baghouses.

Fugitive emissions occur primarily around lead mines and smelters. Lead-laden dusts can be carried off site by winds blowing over tailing piles in mining operations if the tailings are not stabilized by an appropriate cover. Similarly, winds around a smelter can carry lead particles off site from a variety of sources. Control procedures for fugitive emissions around smelters involve complete enclosure of operational areas and improved "housekeeping" techniques. Enclosure improvements could range from fixing broken windows and exhaust hoods, to replacing torn or missing enclosure curtains, or to installing permanent walls around presently unenclosed operations. Such complete enclosures may also require installation of a larger ventilation system, which should be exhausted to the baghouses. Housekeeping improvements involve such things as routine wet-sweeping of open areas, washing tires of vehicles as they leave the plant, and careful work practices.

In addition to the technological solutions mentioned above, full control of emissions depends on a number of human factors. Creative management can play an important role by encouraging employees to adopt work practices that will prevent excessive environmental emissions and maintain workplace safety. If an industrial facility is run haphazardly with improperly trained or unmotivated employees, the result can be dangerous work situations, frequent accidents, low employee morale, and emission problems. If an industry is well run and management demonstrates its concern for workplace and environmental safety by using up-to-date technology and safety equipment and by ensuring that the equipment is maintained properly, the results are fewer emissions and healthier workers.

One tool a plant's management can use to demonstrate its commitment to safe work practices is incentives such as pay bonuses, tickets to a sports event, boxes of steaks, or other forms of employee recognition. Such incentives could be used for the "cleanest crew" or "safety suggestion of the month." Safety can be encouraged also by using large, colourful posters reminding workers of safe work habits. This practice, successful in many different industries, acts as a constant safety reminder. It also has psychological benefits by demonstrating a concern for safety and by helping to brighten up an otherwise dull work environment.

In some cases the necessary condition underlying improvements in emission control is education. If an industry is experiencing emission problems, management and employees may need to be better educated about the serious health and environmental impacts of their poorly controlled operations. This knowledge will help them to understand that their actions can make a difference to their own health and safety as well as to that of people living nearby.

A special control and clean-up situation exists when a lead industry ceases operation at a particular location and the site must be decommissioned. Decomissioning involves carefully controlled dismantling, demolition and site reclamation activities.

Regulatory control over industrial decommissioning in Canada varies from province to province. Ontario is the only province that has any specific guidelines for decommissioning. These guidelines[1] are quite comprehensive but require considerable industry-government cooperation since, as guidelines only, they are not legally enforceable. Other provinces are considering or are actively developing decommissioning guidelines. In the absence of specific guidelines or standards, decommissioning activities are covered by the general prohibitions against environmental pollution contained in various pieces of environmental legislation.

Because of the large potential for environmental contamination when decommissioning occurs, specific, enforceable regulations need to be developed. Successful decommissioning of an industrial facility requires the same kind of careful planning and regulatory control that goes into the original construction of a large facility. The first step in decommissioning is careful and detailed planning of the sequence of activities. Based on the experience of industrial facilities that have already been decommissioned, a crucial part of the planning procedure is a thorough assessment of potential problems, including contingency plans for handling these problems.[2] This type of planning helps to avoid unanticipated "surprises" that can be very costly.

The second step is to determine pre-decommissioning levels of lead in the environment by a detailed sampling and analysis program. Without this type of information, post-decommissioning evaluations of environmental levels is of limited value and may be discredited by the regulatory agency. It is also important to know the specific regulatory requirements that must be met before planning the testing and analysis program.

The primary concern during decommissioning of a lead facility is the control of dust. One method for achieving control is enclosing the entire operation, or portions of the operation, in a "bubble" similar to the bubbles placed over tennis courts in the winter.

Alternatively, dust can be controlled by restricting decommissioning activities to times of low wind speed and by ensuring that water is used to spray down buildings, equipment and soil during excavations. (This water would then have to be filtered before release to the sewers.) If buildings are converted to new use, they will need to be cleaned using specific techniques. These techniques can involve scraping, vacuuming, washing down and cleaning walls, floors, ceilings, sumps and drains, and roofs, with special attention paid to heating and ventilation systems. Special paints may be needed to coat interior surfaces after cleaning.[3] Contaminated equipment will need to be dismantled and either relocated or disposed of properly. Large amounts of contaminated soil may also need to be removed. All materials would have to be shipped to their new location or disposal site in fully enclosed containers.

Industrial decommissioning, in the event of re-location or permanent closure, is an important issue requiring ongoing regulatory attention and public scrutiny. With a few notable exceptions, regulatory control over lead *industries* (in contrast to lead in consumer products) in Canada is generally quite effective. Sustained public pressure directed at individual industries and the government has been highly influential in bringing about this high degree of control.

"Turning Off the Automobile Tap"

The normal operation of an automobile has three potential lead-emission taps that can be turned off or reduced to a much smaller drip. They are exhaust fumes when leaded gasoline is used, disposal of used motor oil from a vehicle operated with leaded gasoline, and disposal of used batteries.

In some high-traffic areas, "mopping up" actions due to exhaust fumes may be necessary (e.g., replacing contaminated soil). However, leaded gasoline use in automobiles is a situation where we will eventually have to "turn off the tap." The first step is bringing it down to a trickle by sharply reducing the amount of lead in leaded gasoline. Lead in gasoline has been phased down or removed in most industrialized countries as of the mid-1980s although some countries, including Canada, are proceeding very slowly.

The second major automobile-related tap is the disposal of used motor oil. Since used motor oil from vehicles running on leaded gasoline can contain up to 10,000 ppm lead,[4] its proper disposal is a matter of some importance.

Many automobile repair garages have holding tanks in which they store waste oil which they sell to re-refiners. Although in Canada in 1985, about 900 million litres of lubricating oil were sold, only about 100 million litres of waste oil were re-refined.[5] Some of the remaining waste oil is typically used to oil dusty roads (which introduces some further lead into the near-roadway environment), but much more of the rest of it, along with the lead it contains, simply enters the environment by one route or another.

If you change your car's engine oil yourself, you should not put the waste oil down the sewer or out with household garbage. Waste oil deposited in storm sewers is simply washed into a body of water (unless the storm sewer system is connected to the sanitary sewer system). The oil portion of waste oil that ends up in a sewage treatment plant is normally broken down satisfactorily, but the lead portion is not. The lead ends up in the sludge at the bottom of the tanks, which is burned, sold to farmers as fertilizer, or placed in local disposal landfills. Any of these alternatives may eventually result in human lead exposure. Putting waste oil out with household garbage means that it probably will end up in a landfill where it may or may not contaminate local groundwater.

Waste engine oil collection depots are a simple and valuable way to recover non-renewable resources. These depots are usually just automobile garages that have their own storage tanks and have agreed to take in waste oil from private citizens. Such networks can be found in many parts of the country. If there is not one in your area, you or your local environmental organization can probably help to get one started. Talk to the local garages or write directly to the refining industry. Governmental agencies also can help. For example, the City of North York in Ontario has sent out a flyer in the municipal water bill describing the facts about waste-oil disposal including a list of addresses of waste oil depots in that city.

The third automobile-related lead source is the disposal of used batteries. Automobile batteries are heavy because of the large amount of lead they contain. Since the lead retains its value even after the battery wears out, many automotive parts distributors or garages will purchase old batteries, which they sell to scrap dealers who in turn sell them to secondary lead smelters.

Although most people turn in their old batteries when a new one is installed, some people install their own new batteries. These people should realize that their old batteries are hazardous waste and should not be discarded with household garbage. If you have an old car battery that you need to get rid of, find a scrap or automotive dealer who will buy it or take it off your hands.

A final source of lead emissions related to automobiles is used engine parts from cars that used leaded gasoline. If these old engine parts are melted down for their metal content in foundries, they can be a source of both occupational-lead exposure and environmental lead contamination. Ignorance about the possibility of lead contamination of old engine parts can lead to serious cases of lead poisoning.

Other lead-emission sources that should be turned off include a variety of household sources (including leaded paint and lead plumbing in older homes with soft, acidic water) that are discussed in Chapter 3.

"Mopping Up the Floor"

In areas where dangerous levels of lead have accumulated, it is necessary to start "mopping up the floor" at the same time as sources are being controlled. The primary location for lead accumulation is soil, although street dust and house dust are also places where significant amounts of accumulation can occur. Since lead is persistent, the current amounts of lead in soil will continue to contaminate root crops or playing children for many years. One prominent environmental scientist has predicted that it will be "...hundreds to thousands of years before there is a significant reduction in lead levels in some soils."[6] The speed of reduction depends on the soil type (i.e., faster reduction in sand and loam, slower in clay) and whether or not the soil is disturbed. Plowing or gardening can result in either moving the lead to deeper layers at a faster rate (a desirable result) or freeing it for wider distribution by breezes or on shoes, animals, or machines (an undesirable result). The slowest rate of reduction will probably occur in undisturbed park land.

Because lead is persistent, because soil is the final resting place for most lead emissions, and because soil is a significant exposure pathway, it has become important to establish a soil-lead level that represents excessive contamination. The excessive contamination level is usually called a soil removal guideline, and is usually applied only in areas to which children have access (e.g., school yards, parks, playgrounds, or yards of private homes).

The beginning of an increase in blood-lead levels in children has been shown to occur when the soil-lead level is between 500 and 1000 ppm.[7] Over 1000 ppm, the blood-lead increase can become larger. Since there is typically a 13% error factor in soil-lead measurements, a soil sample analyzed at 870 ppm could have come from soil that actually had 1000 ppm lead. Based on these facts, we recommend that soil with lead levels of 870 parts per million or more, located in areas to which children have access, be removed and replaced with clean soil. Soil-lead levels above 870 ppm will usually be found only in areas surrounding lead industries and along some heavily-travelled roadways. In addition, when land use changes, soil testing is important to ensure that areas that have been contaminated in the past are not converted to child-accessible areas without consideration of the possible need for soil replacement.

Soil removal involves the excavation of the top 15 centimetres (six inches) of soil, its removal to a specially designated landfill site, and replacement with clean fill and sod. In 1985, estimated removal and replacement costs for a small, urban, residential property were roughly $4000.[8]

Instead of removing the soil to a dump and replacing it with uncontaminated soil, there are ways of separating the lead particles from the soil particles (i.e., "de-leading" the soil) and putting the soil particles back. Techniques for cleaning soil of lead and other contaminants have been developed by several engineering firms in The Netherlands.

Depending on the technique used and the type of soil, anywhere from 30% to 90% of the lead can be removed.[9] The only way to know how much lead can be removed in a particular situation is to test the various methods on samples of the polluted soil. If an adequate method can be found, this soil-cleaning technique represents a permanent solution to contaminant removal and soil preservation and greatly reduces the volume of contaminated material needing further handling. These soil de-leading techniques would probably not be practical for dealing with contamination in residential yards but would be more feasible for large-scale, highly-polluted excavations, such as might arise during industrial decommissioning activites. At present, de-leading equipment for soil does not appear to be commercially available in North America.

Lead also tends to accumulate to high levels in street dust, particularly in cities. With the removal of lead from gasoline, regular street washing will continue to gradually wash this lead into local water bodies (where most of it will end up eventually in bottom sediments) or into sewage sludge.

Other areas needing further clean-up include snow dump sites and leaded paint removal from metal structures such as bridges. Because of the presence of lead from automobile exhausts in snow on streets and roads, snow dump sites should be monitored for lead levels in the soil and, if necessary, considered as contaminated areas with restricted access. The removal of old leaded paint from bridges has been recognized as an occupational hazard but has not usually been considered as local environmental contamination. Especially during a windy period, any people living near bridge paint-removal activities should be advised of the possibility of increased lead exposure and of actions they can take to reduce their exposure.

Once adequate controls have been placed on sources of lead emissions and clean-up actions have removed contaminated soils, the waiting game begins. Even with great reductions in new environmental inputs of lead, it will take many years for significant reductions to occur in foods, soils, and dust. However, over a very long period of time, lead now in the environment will gradually be buried so deeply in soil and sediments that environmental contamination and human exposure problems due to lead will be only something to read about in history books.

Making it Cleaner for the Next Generation

It may be that the greatest tragedy of the present lead-pollution situation is more societal than personal. Without denying the fact that a child with preventable learning disabilities has serious and life-disrupting problems, the risks and disruptions to society may be even greater. In addition to the economic burden of higher costs for health care discussed earlier, widespread contamination of a society with relatively low levels of lead can decrease its vitality, intelligence, and ability to survive. Now that we know these things, we can take actions now that will ensure lower lead levels for the next generation.

An individual citizen actually has more power to bring about changes in the system than he or she may realize. Since most people do not take the time to write politicians about matters that concern them, each letter received by a politician is considered to represent the opinion of a large number of people. A hundred or so individually written letters from different places around the country are an indication that the topic under discussion is of real concern to many people. Whatever common view is expressed in these letters will be given considerable weight. It seems like such a simple thing — just a piece of paper, an envelope, a pen, and a few minutes of time — and yet so few of us do it. If you care about lead risks for your children and grandchildren, write your elected representative and tell him or her what you think. Some major points of national concern are to get the lead out of gasoline and to prohibit further canning of food in cans with lead soldered seams.

Talk to local politicians about changes that could be made to reduce lead levels in your city or neighbourhood. Maybe there is an industry that is releasing lead into the environment. Maybe some soil in playgrounds has high lead levels and needs to be replaced. Maybe some schools and playgrounds are located too close to major traffic routes.

Get together with your friends and neighbours and investigate lead levels and possible lead risks in your area. If you live near a lead industry or high traffic area, find out what the environmental lead levels are and find out how you can receive routine reports of these levels. Your community group might want to set up a "watchdog" committee or person to receive and evaluate these reports and to notify the group if lead increases occur. Get in touch with the industry, let them know you are concerned, and investigate ways you can work together to improve the situation. Your group may want to begin lobbying for better controls on lead or for local clean-up actions.

Educate yourself and your friends on lead issues. Discuss the pros and cons of using unleaded gasoline including reduced automobile maintenance costs. Start a "Tell Two" chain where each person tells two other people about how and why we should reduce lead risks. Realize your power to help get lead out of the environment and out of our lives.

Appendix A

Dying of Lead

The following material is excerpted from a transcript of the January 29, 1974 "As It Happens" CBC Radio Public Affairs programme. At that time an active controversy on lead exposures, health problems,government and industrial responsibilities was occurring.

In the midst of transmitting this programme, "As It Happens" was served an injunction by the Ontario Supreme Court against broadcasting portions of the programme. The injunction had been sought by two Toronto secondary lead smelters. The programme was aired with "bleeps" indicating where material was being deleted under the terms of the injunction.

Harry Brown: ...Right now, in North America, at least a quarter of a million children risk brain damage, mental retardation, and death...from lead poisoning. Each year hundreds of thousands of tons of poisonous lead gets into the air and into our food. All this could be prevented.

From C.B.C. Toronto, "As It Happens" presents a special programme, broadcast simultaneously in the United States. This is: *Dying of Lead*.

Barbara Frum: In Boston, Dr. Herbert Needleman, at the Children's Hospital Medical Centre...

Dr. Needleman: The symptoms of lead poisoning are so vague that many cases of lead poisoning are never picked up. So I, and a number of other people working in the field, are convinced that this is a major health disaster.

Harry Brown: In Toronto, Les Cole lives near the Canada Metal lead plant. Canada Metal for years has been polluting its neighbourhood with lead. Les Cole's five year old son was hospitalized last month.

Mr. Cole: If somebody hadn't come around and asked us about having the children done for the lead test, we wouldn't have known anything about it. His reading was high enough that they took him into the hospital. If nothing had been done, what would it ... how would it affect him later? I mean, if it's left for a couple of years it might have ended up with somebody that's retarded, and that's, you know, that's the worst part of it.

Barbara Frum: In Ottawa, Dr. A. B. Morrison is head of the federal government's Health Protection Branch...

Dr. Morrison: Lead is widely recognized by public health authorities as one of the most dangerous elements to which man is exposed. And that is indicated by the wide range of toxic effects that are associated with lead poisoning. They include things like colic, anemia, convulsions, irritability, blindness, muscular incoordination, sterility, stillbirths, mental retardation, miscarriage — just a tremendously broad range of toxic effects. Some of them, of course, extremely bad.

****

Barbara Frum: In Toronto there are two secondary smelters, which have caused people who live around them a good deal of trouble. Walter Lachocki lives near the plant owned by Toronto Refiners and Smelters. The dirt and dust which comes down on Mr. Lachocki's house contains an amount of lead about 140 times as high as anywhere else in the city...
Mr. Lachocki: I didn't suspect anything was really wrong as far as the high content of lead went in the dust... Dr. Fitch came down from Air Management and he asked if we'd mind having our blood samples taken for high content of lead. At that time I didn't really know what was what.

While Dr. Fitch was here I told him I was concerned about the drums of arsenic that they had in the yard — they're piled up in 200 pound drums all along here — and I said I was concerned about that. And his answer to me was that he wouldn't be too concerned about the arsenic because he'd sooner be poisoned by arsenic than lead. So at that time I got... I sort of started worrying.

He took our lead levels, and in a few weeks' time he said, "Well, they're sort of high. We'll be around in six months' time; they should go down." And of course Air Management all along still wouldn't admit that anything was wrong. And even to this day they don't admit anything was wrong.

Their idea of an immediate health hazard seems to be that if you bring them a corpse, then they might do something. But even then, when we had a meeting with Dr. Fitch, later, when there was a group of doctors at the Hospital for Sick Children, and I inquired that I was concerned that there were five cases of lead poisoning on this street already, and I was concerned — well, his answer to that was that he could bring experts in to prove that there was no such thing as lead poisoning on the block. So I thought, good grief, if that's his attitude, you know, this is a pretty unsafe outfit to deal with.

****

Harry Brown: Morris Kaufman is president of Toronto Refiners and Smelters...
Mr. Kaufman: I don't feel that we have been offending in any way, shape, or form. And if I did, I think we would have shut the plant down years ago.
Barbara Frum: The problem at Toronto Refiners is an old one, yet it has proved very difficult to bring it to public attention. One man who has tried is Dr. David Parkinson, chief of the Metabolic Ward at the Hospital for Sick Children in Toronto. But just listen to what happened to Dr. Parkinson.
Harry Brown: In February 1973 a public meeting was held to plan lead testing in the area of Toronto Refiners. Officials from the provincial Departments of Health and the Environment were there, and they tried to tell the residents that Dr. Parkinson didn't even know what lead poisoning was.
Barbara Frum: Back in 1965, a four year old girl from the Toronto Refiners neighbourhood had been brought into the hospital in a coma. She was critically lead poisoned. Her three brothers and a neighbour child were also treated. The little girl was badly damaged and spent the rest of her life in a mental institution.
Harry Brown: The provincial officials claimed that these weren't lead poisoning cases at all. They tried to tell the neighbourhood residents that Dr. Parkinson was incompetent.
Barbara Frum: Next, the Director of Public Health for Ontario met an official of the Hospital for Sick Children and

complained about Dr. Parkinson. We went to Dr. Parkinson's office to ask him why...

Dr. Parkinson: Well, his concern was two-fold. First of all, he was concerned that I was unnecessarily stirring up emotions and fear in the community, and secondly, that I was, you know, incompetent. That I didn't know how to make diagnoses of lead poisoning and therefore I had no grounds for making any comments about the problem of lead exposure.

Harry Brown: What makes this situation particularly troubling is that the Ministry of Health eventually had to admit that these cases *were* lead poisoning. The question is, why did they try to shut Dr. Parkinson up?

In the studio with us for this evening's programme is Dr. Samuel Epstein, professor of Environmental Health, at Case Western Reserve University School of Medicine in Cleveland. Dr. Epstein is a specialist in pathology and toxicology, the author of more than 130 scientific papers and books, and for the past ten years, has been Chairman of the Air Pollution Control Association's Committee on Biological Effects.

****

Harry Brown: Across town from Toronto Refiners, lead rains down on the neighbours of the Canada Metal plant, too. So far, 11 children from the area have been hospitalized...

****

Harry Brown: The Canada Metal Company is owned jointly by a huge American firm, National Lead, and by Cominco, a subsidiary of Canadian Pacific.

At one point, Cominco's vice president paid a call on the vice chairman of The Hospital for Sick Children, to question the "propriety" of Dr. Parkinson's interest in this issue. Dr. Parkinson thought this was rather, to quote him, "crass" but after all, the provincial Ministry of Health had tried the same thing.

Barbara Frum: Meanwhile, the companies have tried to defend themselves by calling in experts to testify that the lead poisoning is not their responsibility. Ian Outerbridge, of the prestigious Toronto law firm of Thomson Rogers represents both companies...

Mr. Outerbridge: Nobody wants to hear the truth. We originally brought over the best expert we could find in Canada and he was tossed out by the Board of Health because he was working for the Ethyl Corporation.

We then brought in a pediatrician from Chicago, Dr. Henrietta Sachs. And for some reason, she was biased, she was from the United States.

There had to be an expert in this country. Well, there wasn't, so we brought in the best man in the world, a Dr. Barltrop from England, and he was regarded as a whitewash. I don't think really people want to hear the truth.

Harry Brown: The truth. Well, let's see. On October 26th, the Ontario Ministry of the Environment issued a stop-work order against Canada Metal, on the basis of a laboratory report of high lead levels in the blood of three neighbourhood residents. Four days later the stop-work order was set aside by a judge who had listened to Dr. Henrietta Sach — the Dr. Sachs just referred to — who was brought to Toronto by the company to testify. Well, Dr. Sachs testified that the plant just couldn't be guilty.

Barbara Frum: The Ministry of the Environment didn't call any expert to testify for its side at all. And in fact, didn't even show Dr. Sachs its own data on the case.

Harry Brown: The day after the court

hearing we called Dr. Sachs in Chicago, and told her about the data which we'd had for some time.

INTERRUPTION #1 — BLEEPS

Male Narrator: We are interrupting this broadcast. "As It Happens" has just been served, while in the process of transmitting this documentary, with an Injunction from the Ontario Supreme Court against broadcasting portions of it any further. The Injunction was sought by The Canada Metal Company, Limited and Toronto Refiners and Smelters Company Limited and was granted this evening by Mr. Justice Wilson of the Ontario Supreme Court. Therefore, we are legally prevented from proceeding with parts of the documentary entitled ""Dying of Lead." Here is the text of the injunction as handed to the corporation just moments ago.

IN THE SUPREME COURT OF ONTARIO, BETWEEN, THE CANADA METAL COMPANY, LIMITED and TORONTO REFINERS AND SMELTERS COMPANY LIMITED, Plaintiffs — and — CANADIAN BROADCASTING CORPORATION, MARK STAROWICZ AND MAX ALLEN, Defendants

NOTICE OF MOTION

TAKE NOTICE that the Court will be moved on behalf of the plaintiffs at Osgoode Hall in the City of Toronto on Tuesday, the 29th day of January, 1974, at such time as may by the Court be appointed for an interim order, effective for a period of ten days from its date, restraining the defendants and their officers, servants, employees and agents and any other person with knowledge of such order from broadcasting or otherwise disseminating and from advertising or otherwise publicizing a one hour special program prepared with respect to lead poisoning and in particular those aspects thereof alleging or implying that the plaintiffs have brought misleadingly favourable medical evidence and have concealed material evidence from medical experts, or for such other order as may seem just.

AND TAKE NOTICE that in support of such application will be read the Affidavits of Michael Sigel and Carleton Smith, filed, and such further and other material as counsel may advise. DATED at Toronto this 29th day of January, 1974. As It Happens is complying with the terms of the Injunction and is at this moment excising portions of the documentary.

****

BLEEPS — END OF INTERRUPTION #1

Barbara Frum: The Ministry of the Environment has never contacted Dr. Sachs with their data. They've never asked her opinion of it. They have no plans to take Canada Metal to Court again. They have no plans for another stop work order.

Harry Brown: The other expert brought in by the lawyers was Dr. Donald Barltrop of London, England.

Dr. Barltrop: Just saying that someone living near to the industry has a high blood level, begs the question of where the lead came from and what the mode of transmission was. What do we mean by "environmental lead poisoning?" I think we must be very careful before taking an isolated set of samples in relation to the particular plant and necessarily attributing the values obtained to the activities of that plant. It may be that they are related, it may not.

Harry Brown: Dr. Epstein, what has been your experience with the use of experts in cases like this?

INTERRUPTION #2 — BLEEPS

****

Dr. Epstein: You have, I'm sure, witnessed the phenomenon of the growth of the public interest movement in the States, and to a lesser extent in this country. The public interest movement is an expression of gross disillusionment of the public, and of Congress, with the so-called "expert," the man who is paid to say something for his own personal gain or profit, or for the gain or profit of the industry to which he belongs. And the emphasis is always to maximize the profit, even if it means maximal risks and minimum benefit to the public and to the consumer.

****

Barbara Frum: Dr. Epstein, if all this lead is so dangerous, then why don't we put a stop to it? I feel like a guinea pig...

Dr. Epstein: There is a growing body of evidence, both clinical and experimental, that subclinical brain damage and other subtoxic effects occur from exposure to levels of lead which have previously been considered safe.

The effects of this in terms of human health, in terms of mass impact on society — in terms of health and in terms of economics — are so massive, that to allow this major experiment in human toxicology to proceed for the economic interest of a narrow segment of society is something which is totally reprehensible.

****

A copy of the complete transcript as it was aired is available in the Community Lead Impact Study's resource collection in Sanderson Public Library, Toronto.

Appendix B

Occupations that may Involve Exposure to Lead

Ammunition makers and users
Auto mechanics
Automobile workers
Babbitters
Battery makers
Bookbinders
Bottle cap makers
Brass founders
Brass polishers
Braziers
Brick burners
Bronzers
Brush makers
Cable makers
Cable splicers
Canners
Cartridge makers
Ceramic makers
Chemical equipment makers
Chippers
Cutlery makers
Demolition workers
Dental technicians
Diamond polishers
Dye makers
Electronic device makers
Electroplaters
Electrotypers
Emery wheel makers
Enamel burners
Enamelers
Enamel makers
Farmers
File cutters
Filers
Flower makers, artificial
Foundry molders
Galvanizers
Gasoline blenders
Glass makers
Glass polishers
Gold refiners
Gun barrel browners
Incandescent lamp makers
Insecticide makers
Insecticide users
Japan makers
Japanners
Jewelers
Junk metal refiners
Lacquer makers
Lead burners
Lead counterweight makers
Lead flooring makers

Lead foil makers
Lead mill workers
Lead miners
Lead pipe makers
Lead salt makers
Lead shield makers
Lead smelter workers
Lead stearate makers
Lead workers, other
Linoleum makers
Linotypers
Lithographers
Match makers
Metal burners
Metal cutters
Metal grinders
Metal miners
Metal polishers
Mirror silverers
Motor fuel blenders
Musical instrument makers
Painters
Paint makers
Paint pigment makers
Patent leather makers
Pearl makers, imitation
Pipe fitters
Plastic workers
Plumbers
Pottery glaze mixers
Pottery workers
Printers
Putty makers
Riveters
Roofers
Rubber buffers
Rubber makers
Scrap metal workers
Sheet metal workers
Shellac makers
Ship breakers
Shipbuilders
Shoe stainers
Shoe makers
Solderers
Solder makers
Steel engravers
Stereotypers
Tannery workers
Telephone repairers
Temperers
Tetraethyl lead workers
Tetramethyl lead workers
Textile makers
Tile makers
Tin foil makers
Tinners
Traffic work
Type founders
Typesetters
Varnish makers
Wallpaper printers
Welders
Wire patenters
Zinc mill workers
Zinc smelter chargers

Source: *Adapted from lists developed by the United States Department of Health, Education and Welfare/Public Health Service, National Institute for Occupational Safety and Health; and related sources.*

Appendix C

Home Lead Inventory

Use this list of questions to find possible sources of lead in and around your home. If you think you or your children have had considerable exposure to lead from sources such as these, ask your doctor for a blood-lead test.

Part I: Children

Do your children ever play with or eat paint chips or plaster?
Do your children chew or suck on painted toys, furniture, or outdoor play equipment?
Do your children frequently put their fingers, toys, or other small objects in their mouths?
Do your children eat soil or other non-food items?
Do your children play in old or abandoned housing or on vacant lots where old housing has been torn down?
Do your children play with discarded battery casings or other lead-containing objects?
Do your children have any hobbies that use lead metal or lead solder (e.g., electronics kits, metal work, stained glass)?
Do your children play or go to school near a lead-related industry or heavily travelled road?
Do your children play where the soil could be contaminated with paint chips from old buildings?
Do you allow your children to burn lead-containing materials such as old, painted wood?

Part II: House and Yard

Is your house or apartment building painted with high-lead (pre- 1977) paint?
Is there paint or plaster chipping or flaking off the walls or woodwork?
Have you redecorated or removed old paint recently?
Are you planning to redecorate or remove old paint in the future?
Do you have lead plumbing pipes and soft water?
Do you use glasses or dishes decorated with painted figures?
Are there lead objects around the house, such as toy soldiers, jewelry, ammunition, beads, fishing sinkers, or old toothpaste and ointment tubes?
Do you serve or store foods or drinks in lead-glazed ceramic dishes that are badly fired or cracked?

Do you use pewter containers for foods or drinks?
Does your house get dusty very quickly?
Is there any lead-based paint in areas where children eat, play, or sleep?
Is there lead-based paint on outdoor furniture or children's toys or furniture?
Do your children eat mostly processed foods (in contrast to fresh foods)?
Do you always wash your children's hands well before meals or snacks?
Have you asked these kinds of questions about any other place where your children regularly spend time, such as a babysitter's house or daycare centre?

Part III: Family and Neighbourhood

Is anyone in your family pregnant or planning on ever becoming pregnant (the fetus is very sensitive to lead in the mother's blood)?
Does anyone in your family work in a lead or lead-related industry (see list in Appendix B)?
Does anyone in your family have a hobby that uses lead (e.g., painting, potting, soldering, stained glass working, or doing graphic arts)?
Does anyone in your family drink bootleg whiskey that might have been produced in a lead-containing still or homemade wine that has been in contact with any lead?
Is there a lead smelter, lead-using industry, or scrap metal salvage yard in your neighbourhood?
Do you live near a road with heavy traffic?
Has anyone in your family or neighbourhood had a high blood-lead level?
Is there any peeling paint on fences, buildings, or garages in your neighbourhood?
Are there vacant lots nearby where old car batteries or other lead-containing trash may have been discarded?

Part IV: Diet

Does everyone in your family, especially the children, consume an adequate amount of calcium and iron?
Do you eat food from a garden located near a lead industry or heavily travelled road?
Do you use a lot of canned foods that are not in lead-free cans?
Do you ever store food or juices in an open can?
Do you use cans that are dented or otherwise damaged?
Do you eat foods that are high in lead? (Check the table below.)

Table C.1

Lead Levels of Some Foods in Canada

Uncanned Food	**Lead (ppb)**
Most dairy products	20-50
Cheese	240
Beef, pork, lamb, poultry	50-100
Eggs	50
Great Lakes fish	200-600
Cereals, bread, rice, flour	40-80
Potatoes, yams	60-80
Vegetables: leafy, root, tomatoes, beans, peas	30-70
Citrus fruits and juice	30
Apples, apple products	140
Other fruits, berries	20-50
Sugars, sweeteners, candy	50-60
Molasses	410
Chocolate bars	100
Coffee, tea	30
Infant foods:	
Liquid formula	30
Powdered formula	10
Strained meats	30
Strained vegetables	40
Juices, drinks, strained fruits	50
Cereals	90

Effect of Canning on the Above Values

Food Group	**Lead Increase**
Vegetables: leafy, root, potatoes	2-5 times
Tomatoes	6 times
Fruits	1.2-7 times
Fish	5 times

Sources: *Adapted from Nutrition Foundation, Inc., "Assessment of the Safety of Lead and Lead Salts on Food: A report of the Nutrition Foundation's Expert Advisory Committee," Washington, D.C., The Nutrition Foundation, 1982; and Great Lakes fish levels from J. O. Nriagu, "Lead Contamination of the Canadian Environment," unpublished report to the Royal Society of Canada's Commission on Lead in The Environment, 1985.*

Appendix D

Information for Soil Lead Testing

The following information was provided by provincial environmental departments.

British Columbia
Private citizens may submit soil for analysis to the Environmental Laboratory, Ministry of the Environment, 3650 Wesbrook Crescent, Vancouver, British Columbia V6S 2L2, Phone 604-660-1685.

The following metals can be analyzed: aluminum; antimony; arsenic; barium; boron; cadmium; calcium; chromium; cobalt; copper; iron; lead; magnesium; manganese; molybdenum; nickel; silver; uranium; vanadium; zinc

The current charge for the service is $86.00

Alberta
Contact the Director of Pollution Control 413-427-6270 (Edmonton)

Saskatchewan
A soil-testing service is not routinely available, but the Saskatchewan Research Council labaoratories in Regina can do analyses for heavy metals for a fee.

Manitoba
Private citizens must contact the office of Environmental Management, Department of Workplace Safety and Health, Building 2, 139 Tuxedo Avenue, Winnipeg, Manitoba R3N 0H6 and explain his/her requests. The office will assess whether the request for analyses is warranted by checking on already existing information. If analyses are considered justified, samples are taken by departmental staff. The service is free to Manitobans.

The following metals can be analyzed: arsenic; cadmium; copper; lead; mercury; nickel; and zinc.

Other, more unusual metals can be analyzed providing method development and quality control studies are carried out in advance in the department's laboratory.

Ontario
The Ministry of the Environment will test the lead content of soil for free. Call 416-965-4516 (Toronto). The ministry's testing usually provides results of arsenic, cadmium, copper, antimony and zinc, in addition to lead.

Quebec
This province did not provide the authors with any information.

New Brunswick
The government does not provide free soil testing to private citizens except in cases where a spill of contaminants is being investigated. Apparently, there are a limited number of consultants and laboratories in New Brunswick equipped to take proper samples and provide analyses of soil for lead and heavy metals. Citizens who suspect they may have a problem relating to heavy metals in soil should contact the Department of Municipal Affairs and Environment for advice through the Director, Pollution Control Branch, 506-453-2861 (Fredericton) or by writing to him/her at Box 6000, Fredericton, New Brunswick E3B 5H1.

Nova Scotia
Soil sample analysis for heavy metals is available by contacting the Environmental Chemistry Laboratory, Victoria General Hospital in Halifax, 902-428-3464. There is a fee for the service.

Prince Edward Island
This province did not provide the authors with any information and does not have an environmental ministry.

Newfoundland and Labrador
Soil testing is done on a cost recovery basis for soil conditions such as acidity and micro-nutrients but lead and other heavy metals are not checked on a routine basis. Information can be obtained from the Soil and Plant Laboratory, Department of Rural, Agricultural and Northern Development, 705-576-3845 (St. John's).

United States
State departments of agriculture or local departments of public health usually can provide information on soil-testing possibilities in your local area.

Appendix E

An Environmental Bill of Rights

The primary purpose of an environmental bill of rights is to state clearly and explicitly that the citizen has a right to a healthy environment. In order to provide for this right, government has a duty to protect the environment from degradation. This duty means that the government's first priority on accepting any proposal (private or public) is the protection of the environment. A system of mechanisms and procedures needs to be available to both citizens and the government to provide a basis for backing up these rights.

The environment can be broadly defined in terms of all of the following:

1. Air, land, or water
2. Plant and animal life, including human
3. The social, economic, and cultural conditions that influence the life of people or a community
4. Any building, structure, machine, or other device or thing made by man
5. Any solid, liquid, gas, odour, heat, sound, vibration, or radiation resulting directly or indirectly from the activities of humans

An environmental bill of rights should operate at both the federal and provincial levels. The following statements cover some of the basic elements of such a bill.

1. Environmental quality is an unalienable right, for without an environment capable of supporting the human race, all other rights are useless. This would provide protection from pollution rather than prosecution after environmental degradation has occurred.
2. Any member of the public should have the right to enforce the law and uphold his/her common law rights to protect the environment — a right known as "standing." This right would allow citizens to use the law to defend the environment in courts, tribunals, hearings, etc.
3. Before beginning any undertakings that might endanger the environment, environmental impact studies should be made. Ontario has an *Environmental Impact Assessment Act* with many good features. However, it is rarely used since the government frequently exercises its power to exempt any undertaking (public or private) from assessment. Exemptions should not occur.
4. Access to information, both governmental and corporate, that concerns environmental matters should be a fundamental and necessary right.
5. Citizens should have the right to participate in the setting of standards of

environmental quality (e.g., acceptable levels of pollution) and the right to demand review of these standards when there is new technology or information.

6. There should be an environmental ombudsperson to protect the rights of natural objects and unborn generations, and to advise, review, and report on the condition of the environment.
7. Any citizen should be able to sue on behalf of other similarly aggrieved citizens (a class action) for damages resulting from environmental destruction.
8. Citizens should have the right to defend the environment at a reasonable cost. There should be a mechanism to provide funds for legal and scientific consultation to citizens using the courts or hearing panels.
9. Legislative actions should be implemented by concrete guidelines that would permit review of an agency's actions by the ombudsperson or the courts.
10. Courts should be able to review agency's actions made in the public interest for their wisdom, adequacy and fairness.
11. The burden of proof of environmental safety of a new product, substance, machine, or process should be with whoever introduces it into use.

Source: *Reproduced with permission from "The Hazardous Waste Educational Kit," Federation of Ontario Naturalists, 1985.*

Appendix F

Measurement Units

Basic Units and Their Abbreviations

kg	kilogram	µg/g	micrograms per gram
g	gram	µg/m^3	micrograms per cubic metre
mg	milligram	mg/m^2	milligrams per square metre
µg	microgram	mg/g	milligrams per gram
L	litre	µg/dl	micrograms per decilitre
dl	decilitre	µg/L	micrograms per litre
ml	millilitre	µg/kg	micrograms per kilogram
m	metre	µmol/L	micromoles per litre
t	tonne	µg/100ml	micrograms per 100 millilitres
lbs.	pounds	µg/kg/d	micrograms per kg per day
m^3	cubic metre	ppm	parts per million
		ppb	parts per billion

Equivalent Measures

1000 g	= 1 kg	1%	= 10,000 ppm
1000 mg	= 1 g	1 µg/g	= 1 mg/kg= 1 ppm
1000 µg	= 1 mg	1 µg/kg	= 1 ppb
1000 ml	= 1 L	1000 µg/kg	= 1 µg/g
1000 ppb	= 1 ppm	2000 lbs.	= 1 ton
1000 kg	= 1 tonne	= 2204.6 lbs.	

Blood-lead Equivalent Measures

Blood Lead	μmol/L x 20.7	= μg/dl
EP* Lead	μmol/L x 56.5	= μg/dl
Blood Lead	μg/dl x 0.048	= μmol/L
EP* Lead	μg/dl x 0.018	= μmol/L

1 μg/dl = μg/100 ml = 0.1 ppm = 100 ppb

Note that μmol/L is the official metric system (SI) measure.

* EP = erythrocyte protoporphyrin, a substance that does not itself contain lead but that is an indirect measure of lead in blood.

Lead Measurements

There are two ways to look at lead measurements in the body. The first way is how much is taken in each day. This measurement is called "intake." An adult daily intake of 250 micrograms can cause serious lead poisoning. The second way to look at lead measurements is how much is in the blood. This measurement is called "blood-lead level." A blood-lead level of 80 to 120 micrograms can cause death. The amount of lead in the blood is determined by a combination of factors including present daily intake, the amount that is excreted, and the amount that was already present in the body from previous intakes. These previous intake amounts could have been stored in blood, soft tissues (e.g., organs, muscles, etc.), or bone.

"Safe" Blood-lead Levels

The medical profession does not like to use the word "safe" in reference to blood-lead levels, because no one really knows how much is safe. They are, however, prepared to say how much is "too much." The words that have been used over the years to describe "too much" lead have included poisonous, dangerous, toxic, excessive, and elevated. The actual blood-lead levels to which these words referred has been as varied as has the terminology, but in general has become lower and lower as more evidence has been gathered on the effects of lead on health. The chart below shows these changes over the years in the "too much" level, as well as the "action" level. The "action" level refers to the blood- lead level at which there is some concern about the amount of lead in the blood even though it isn't high enough to indicate that medical intervention is needed. At the "action" (i.e., intervention) level, there might be some investigation of possible exposure sources or there might be counselling on changes in lifestyle that could lower the blood-lead level.

Table F.1

"Too Much" and "Action" Blood-lead Levels over the Years as Applied to Children

Year	μg/dl	Terminology	μg/dl	Terminology
pre-1970	80	dangerous	60	intervention
1970	80	poisonous		
1970	60	dangerous		
1973	40	medically undesirable		
1975			30	intervention
1977	35	dangerous		
1978	30	toxic		
1985	25	elevated	20	intervention or allowable

In 1985, "toxic" is used with a blood-lead level of 25 μg/dl if there is also an EP measure of 35 μg/dl or more.

Sources: *United States Environmental Protection Agency, "Health Implications of Airborne Lead," November, 1973; and Centers for Disease Control, "Preventing Lead Poisoning in Young Children," United States Department of Health and Human Services/Public Health Service, Atlanta, Georgia, 1985, p. 3.*

Appendix G

Blood-lead Level Estimates

PART 1: Source Contribution Table

Estimated Contributions of Various Environmental Sources of Lead to Blood-lead Levels*

Estimated Contributions in ug/dl

Residence	Food	Dust	Air	Water	Total	Actual Blood Lead (μg/dl)
			Children - 2 years old			
Rural	5.92	0.77	0.19	0.10	6.98	8.91
Urban	5.92	5.26	1.04	0.10	12.32	12.02
ULI	5.92	8.23	1.35	0.10	15.60	16.5, 15.0
			Adults			
Rural	2.18	0.26	0.16	0.03	2.63	N A
Urban	2.18	1.75	0.88	0.03	4.84	10.1
ULI	2.18	2.74	1.15	0.03	6.10	11.6

Key
ULI = residences within about 200 metres of an urban lead industry
N A = not available

* Note that in this and the following two tables the adult values do not include any amounts due to occupational exposure (variable amount), alcohol consumption (about one μg/dl), and smoking (about two μg/dl).[1]

Derivation of the Above Values

1. Child and adult food estimates are based on estimated Canadian food-lead intake in micrograms per day (μg/d) and a food-lead to blood-lead factor used by the United States Environmental Protection Agency. The current daily food-lead intake estimates were adapted from values published by Environment Canada.[2] These values, of 62 μg/d for children age 1 to 3 years and 113 μg/d for adults, are comparable to other Canadian food-lead values from about 1972. In order to reflect the current condition of reduced levels of lead in food, the values of 62 and 113 μg/d have been reduced by 40% to 37 and 68 μg/d.

The basis for this reduction is the fact that over the seven year period that elapsed between two extensive food-lead estimates in the United States (i.e., the period 1976-1980 compared to 1981-1983), there was about a 20% reduction in food-lead levels.[3] In addition, we have assumed a further 20% reduction since current U.S. food estimates (i.e., 23 μg/d for two-year old children and about 40 μg/d for adults[4]), are considerably below the Environment Canada figures and since a further six years has elapsed since the earlier Canadian estimate and the date the blood-lead levels were obtained (i.e., 1972 to 1976 and 1983 to 1985). The 40% reduction in the Canadian food-lead estimates appears to be approximately correct and a conservative reduction since the adult estimate of 68 μg/d is similar to the 52 μg/d 1985 estimate by Health and Welfare Canada for adults.[5]

The child daily intake estimate of 37 μg/d (60% of 62) was multiplied by 0.16, a food intake to blood-lead factor (Ryu[6]) (37 x 0.16 = 5.92). The adult daily intake estimate of 68 (60% of 113) was multiplied by one-fifth the Ryu factor since adults typically absorb only about one-fifth as much gut lead as do children (68 x 0.032 = 2.18).

2. The dust estimates are based on Angle et al, 1984 and Ryu et al, 1983 equations[7] as modified by Canadian dust-lead estimates of 60 ppm and 850 ppm in rural and urban areas, respectively[8] and 1700 ppm in ULI (a conservative estimate from this report based on twice the urban level and falling within the lower range of Nriagu's estimates[9] for street and house dusts near a smelter of 200 to 5000 ppm).

The contribution of non-atmospheric lead dust to blood lead is estimated by:

0.10 x (soil lead in ppm x 0.00681 + house dust in ppm x 0.00718) where our house dust estimate was 2.2 times the soil-lead level.[10]

The contribution of atmospheric lead dust to blood lead was derived by adjusting the blood-lead estimates[11] for various air lead levels by Canadian air-lead levels for rural and urban locations of 0.10 μg/m^3 and 0.54 μg/m^3, respectively[12] and for ULI locations of 0.70 μg/m^3 [13] representing the arithmetic mean of the yearly geometric means for the nearest four stations to two urban lead smelters in Toronto.

Computations for children:
Rural: non-atmospheric
0.10 x (60 x 0.00681 + 132 x 0.00718) = 0.14
: atmospheric (nearest air-lead concentration in USEPA table is 0.25 μg/m^3)
0.10/0.25 x 1.57 (value in table) = 0.63
Total 0.77

Urban: non-atmospheric
0.10 x (850 x 0.00681 + 1870 x 0.00718) = 1.92
: atmospheric (nearest air-lead concentration in USEPA table is 0.50 ug/m^3)
0.54/0.50 x 3.09 (value in table) = 3.34
Total 5.26

ULI: non-atmospheric
0.10 x (1700 x 0.00681 + 3740 x 0.00718) = 3.84
: atmospheric (nearest air-lead concentration in USEPA table is 0.75 ug/m^3)
0.70/0.75 x 4.70 (value in table) = 4.39
Total 8.23

Computation for adults:
No equations have been developed for adults. Therefore, since adults absorb about 10% of ingested dust compared to 30% for children, adults' values were estimated at one-third children's values.

Rural	0.77/3	= 0.26
Urban	5.26/3	= 1.75
ULI	8.23/3	= 2.74

3. The air estimates are based on Canadian air-lead data and on air- to-blood-lead slope estimates of 1.93 for children[14] and 1.64 for adults.[15] The sources for the rural, urban, and ULI air estimates of 0.10, 0.54, and 0.70 were described in the dust factor section above.

Children:

Rural	0.10 x 1.93	= 0.19
Urban	0.54 x 1.93	= 1.04
ULI	0.70 x 1.93	= 1.35

Adults:

Rural	0.10 x 1.64	= 0.16
Urban	0.54 x 1.64	= 0.88
ULI	0.70 x 1.64	= 1.15

4. The water estimates are based on estimated Canadian water-lead intake (μg/day) of 0.6 for children and 1.4 for adults[16] and the water intake to blood-lead factor of 0.16 for children.[17] For adults one-fifth of the 0.16 factor was used since adults absorb about one-fifth as much gut lead as children.

Children:	0.6 x 0.16	= 0.10
Adults:	1.4 x 0.032	= 0.04

5. The actual blood-lead values for rural and urban children were taken from the Ontario Tri-ministry report.[18] The 16.5 μg/dl value for ULI children (N = 7) and the 11.6 value for ULI adults (N = 87) were obtained in a 1985 screening program in the Niagara Neighbourhood smelter area in Toronto.[19] The 15.0 μg/dl value for ULI children (N = 434 including multi-year testing on many children) was obtained from three blood-screening programs from 1983 to 1985 in the South Riverdale smelter area in Toronto.[20] The children were under six years in all screenings, and the adults were 18 years and over. The 10.1 value for urban adults was obtained from a 1981 study on 338 blood donors in Alberta.[21]

Endnotes for Appendix G Part 1.

1. J. Roberts, K. R. Mahaffey, and J. L. Annest, "Blood Lead Levels in General Populations," in K. R. Mahaffey (ed) *Dietary and Environmental Lead: Human Health Effects,* New York, Elsevier, 1985, pp. 355-372.
2. Environment Canada, "Socio-economic Impact Analysis of Lead Phase-down Control Options," Environmental Protection Service, February, 1984, p.21.
3. United States Environmental Protection Agency (hereinafter referred to as USEPA), "Air Quality Criteria for Lead," EPA-600/8- 83-028B, Environmental Criteria and Assessment Office, Research Triangle Park, N.C., 1984, pp. 7-47 to 7-48.
4. R. W. Elias, "Lead Exposures in the Human Environment," in K. R. Mahaffey (ed), op. cit., pp. 79-107.
5. R.W. Dabeka, Health and Welfare Canada, personal communication, April 16, 1986.
6. J. E. Ryu, E. E. Ziegler, S. E. Nelson, S. J. Fomon, "Dietary Intake of Lead and Blood Lead Concentrations in Early Infancy," *Am. J. Dis. Child.,* 1983, 137, pp. 886-891 as cited in USEPA, op. cit., p. 1-134.
7. C. R. Angle, A. Marcus, I-H Cheng, M. S. McIntire, "Omaha Childhood Blood Lead and Environmental Lead: A Linear Total Exposure Model," *Environ. Res.,* 1985, 35; ibid.
8. Environment Canada, op. cit.
9. J. O. Nriagu, "Lead Contamination of the Canadian Environment," unpublished report to the Royal Society of Canada's Commission on Lead in the Environment, 1985.
10. I. Thornton, E. Culbard, S. Moorcroft, J. Watt, M. Wheatley, and M. Thompson,"Metals in Urban Dusts and Soils," *Env. Tech. Letters,* 1985, 6, pp. 137-144.
11. USEPA, op. cit., p. 1-134.
12. Environment Canada, op. cit.
13. Ontario Ministry of the Environment, personal communication, March 11, 1986.
14. C. R. Angle et al, op. cit.
15. USEPA, op. cit., p. 11-104 for air lead levels equal to or less than 3.2 ug/m^3 and blood lead levels equal to or less than 30 ug/dl in adults.

16. Environment Canada, op. cit., p. 21.
17. Ryu, op. cit.
18. C. Duncan, R. A. Kusiak, J. O'Heany, L. F. Smith, L. Spielberg, and J. Smith, "Blood Lead and Associated Risk Factors in Ontario Children, 1984: Summary and Conclusions," Ontario Ministries of Environment, Health, and Labour, 1985, p. 20.
19. D. McKeown, "Capillary Blood Lead Levels: Niagara St. Lead Screening Clinic, July 1985." Letter to B. Wallace, August 26, 1985.
20. E. Ellis, Toronto Department of Public Health, personal communication, May, 1986.
21. J. I. Cheng, "A Blood Lead Survey of Southern Alberta," paper presented to Canadian Public Health Association Conference, Saskatoon, June 26, 1981.

PART 2: Daily Lead Intake and Absorption per Kilogram Body Weight

Intake

Micrograms per kilogram body weight per day for children and adults living in rural, urban, and urban-lead industry (ULI) areas

	Children			Adults		
Source	**Rural**	**Urban**	**ULI***	**Rural**	**Urban**	**ULI***
Food	2.86	2.86	2.86	0.97	0.97	0.97
Dust	0.46	6.54	13.08	0.02	0.24	0.48
Air	0.05	0.25	0.32	0.04	0.04	0.20
Water	0.05	0.05	0.05	0.02	0.02	0.02
TOTAL	3.42	9.70	16.31	1.05	1.27	1.67

* within approximately 200 metres of an urban smelter.
Body weights used: 13 kilograms for a two-year old child and 70 kilograms for an adult.

Source: *Environment Canada, "Socio-economic Impact Analysis of Lead Phase-down Control Options," Environmental Protection Service, February, 1984; D. Ogner, Ontario Ministry of the Environment, personal communication, August, 1985; and R. Elias, United States Environmental Protection Agency, personal communication, July, 1985.*

Absorption

Micrograms per kilogram body weight per day for children and adults living in rural, urban, and urban-lead industry (ULI) areas

	Children			Adults		
Source	**Rural**	**Urban**	**ULI ***	**Rural**	**Urban**	**ULI ***
Dust	0.14	1.96	3.92	0.001	0.02	0.05
Air	0.02	0.10	0.13	0.01	0.06	0.08
Water	0.02	0.02	0.02	0.001	0.001	0.001
TOTAL	1.61	3.51	5.50	0.11	0.18	0.33

* within approximately 200 metres of an urban smelter.
Body weights used: 13 kilograms for a two-year old child and 70 kilograms for an adult.

Source: *Daily intake table adjusted by the following absorption factors: food and water, for children 50% of intake and for adults 10% of intake; dust, for children 30% of intake and for adults 10% of intake; air, 40% of intake.*

PART 3: Actual Blood-lead Levels in an Urban Lead-industry Area*

Year	N	Geometric Average	Standard Deviation
1983	70	15.17	1.37
1984	216	14.57	1.41
1985	148	15.27	1.45

*Results of finger-prick blood-lead screening programs from 1983 to 1985 on children under six years in the South Riverdale area of Toronto where a secondary lead smelter is located.

Source: *E. Ellis, Toronto Department of Public Health, personal communciation, May, 1986.*

Appendix H

Costs and Benefits of Leaded and Unleaded Gasoline

The three different topics related to costs and benefits associated with the use of leaded gasoline are:

1. misfuelling with leaded gasoline in cars that are intended to use unleaded gasoline;
2. phase-down of the lead content of leaded gasoline from 0.29 g/L (the Canadian standard as of 1987) to 0.026 g/L (the U.S. standard as of 1986); and
3. using unleaded gasoline in older cars that were originally designed to use leaded gasoline.

Misfuelling

The following table outlines the various effects on a car's engine resulting from the incorrect use of leaded fuel. Costs are approximate, quoted in U.S. dollars (1984), and do not include any towing or associated expenses that may accompany these repairs.

Phase-down of Lead in Leaded Gasoline

The costs and benefits of reducing the lead content of gasoline from 0.29 g/L to 0.026 g/L were calculated by the United States Environmental Protection Agency. The following table shows the year-by-year costs and monetized benefits of the U.S. phasedown which took effect in a step-wise progression beginning in July, 1985 (0.13 g/L) and January, 1986 (0.026 g/L). The analysis assumed that some misfuelling would continue to occur.

The net benefits to the U.S. over the seven and one half years, excluding the benefits associated with blood pressure- related health effects, are over $8 billion. When the costs of blood pressure-related health effects are included, the net benefits to the U.S. exceed $46 billion. This large figure may in fact justify a complete ban on lead in gasoline by 1988. The level of 0.026 g/L was chosen, however, to reduce the lead content of gasoline as much as possible while avoiding the potential problem of valve seat recession in the small number of vehicles susceptible to it if they use totally lead-free gasoline. All of the cost estimates are conservative underestimates of total savings. In addition, many significant but unquantifiable savings are necessarily omitted. If Canada were to recieve one-tenth of these savings from a similarly swift phase-down, $800 million (excluding blood pressure-related effects) or $4.6 billion (including blood pressure-related effects) would be saved over a similar seven to eight year period.

Using Unleaded Gasoline in Older Cars

Based on extensive testing in the U.S. on several fleets of cars built to use leaded gasoline (and equipped with hardened valve seats or valve seat inserts), the use of unleaded gasoline was found to be entirely acceptable under normal operating conditions. (The use of super unleaded is advisable if driving with regular heavy loads or very high speeds.)

The greater at-pump cost of unleaded fuel is roughly offset in these cars by lower maintenance costs and increased fuel economy.

Most of the tests to determine the benefits of the reduced lead content (0.026 g/L) on maintenance and fuel economy were done on cars that switched from leaded to unleaded gasoline rather than to a very low-lead gasoline such as 0.026 g/L. The maintenance and fuel economy savings were therefore interpolated, but the feasibility and benefits of the switch from leaded to unleaded were proven. Therefore, out-of-pocket costs for the operation of an older car using either leaded or unleaded gasoline are approximately equal when all variables are taken into account.

Endnotes for Table H.1

1. the amount of money, due to misfuelling, in excess of what would normally be spent on maintenance over 80 000 km.
2. most post-1975 cars
3. most post-1981 cars
4. if the price differential remains as in 1986, i.e., three to four times greater than the manufacturing cost difference (see text).

Source for Table H.1 *"Incentives for Proper Usage of Unleaded Fuel," Motor Vehicle Manufacturers' Association, January 11, 1984.*

Table H.1

Effects of Misfuelling

Engine Component	Effect on Component	Engine Performance	Driveability	Fuel Economy	Cost Penalty over 80 000 km[1]
catalytic converter[2]	plugged, causing increased back pressure	reduced	reduced	reduced	$300/replacement
spark plugs	fouled	reduced	reduced	reduced	$35 (frequent changes)
exhaust gas oxygen sensors[3]	deteriorate and/or collect deposits	poor	poor	reduced	$75/replacement
oil and oil filter	heavy sludge forms causing deterioration and filter clogging	earlier need for major engine repairs	shorter oil change intervals		$60 (frequent changes)
EGR Valve (Exhaust Gas Re-circulation)	EGR valve passages can clog		potential problems (spark knock)		$60/replacement
					Total $530/80 000km

Additional unquantifiable considerations include:
Warranty coverage (engine component failure resulting from the incorrect use of leaded fuel may void warranty coverage, and simple methods exist for detecting leaded fuel use).
Air quality (increased emissions of pollutants normally controlled by catalytic converters).
Vehicle value (loss in value due to reductions in performance and driveability).
Other systems (e.g., induction and exhaust affected in many different ways that are difficult to quantify).

Summary in U.S. dollars

fuel cost savings using leaded instead of unleaded gasoline	$130/80 000 km[4]
	minus
cost penalty of using leaded instead of unleaded gasoline	$530/80 000 km
overall cost penalty	$400/80 000 km

The cost savings are approximately four times greater that the fuel savings realized.

Table H.2

Year-by-Year Costs and Monetized Benefits of Final Phase-Down Rule, Assuming Partial Misfuelling (millions of 1983 dollars)

Monetized Benefits	**1985**	**1986**	**1987**	**1988**	**1989**	**1990**	**1991**	**1992**
Children's health effects	223	600	547	502	453	414	369	358
Adult blood pressure	1724	5897	5675	5447	5187	4966	4682	4691
Conventional pollutants	0	222	222	224	226	230	239	248
Maintenance	102	914	859	818	788	767	754	749
Fuel Economy	35	187	170	113	134	139	172	164
Total Monetized Benefits	2084	7821	7474	7105	6788	6517	6216	6211
Total Refining Costs	96	608	558	532	504	471	444	441
Net Benefits	1988	7213	6916	6573	6284	6045	5772	5770
Net Benefits Excluding Blood Pressure	264	1316	1241	1125	1096	1079	1090	1079

Source: *Adapted from "Costs and Benefits of Reducing Lead in Gasoline: Final Regulatory Impact Analysis," EPA-230-05-85-006, Office of Policy Analysis, United States Environmental Protection Agency, Washington, D.C., February, 1985, p. E-12.*

Appendix I

Regulatory Recommendations

Several chapters in this book provide valuable lifestyle suggestions for ensuring that your lead risks are as low as possible. However, we should not be expected to have to continually avoid all of the lead in the environment. As discussed in Chapter 14, it is essential that our society greatly improve its control over the current releases of lead into the environment. Some suggested improvements in lead control are summarized in the following recommendations.

Control Limits

Gasoline

A gasoline lead standard of 0.026 g/L should be adopted in Canada in 1989. (This date provides the same two year lead-time to the industry as was provided in the U.S.) Lead should be eliminated from gasoline as soon as is technically feasible thereafter.

Food

The regulations for the lead content of food should be brought up to date in two ways. First, current regulations need to be reviewed and, in most cases, lowered considerably. Second, more foods should be included on the list of regulated foods.

Lead solder in food cans should be banned.

Dustfall

Ontario's guideline for lead in dustfall around lead-emitting industries needs to be lowered, and made enforceable.

Soil

The soil-removal guideline of 2600 ppm used in Ontario, Alberta, and Manitoba should be lowered at least to 870 ppm for areas accessible to children. If elevated soil-lead levels are found within a 1000 ppm isopleth around an industry, the financial responsibility for removal should be borne equally by that industry and the government. In other areas, the government should assume full financial responsibility.

Paint

The federal lead in paint regulation in Canada should be lowered to 600 ppm similar to the U.S. regulation.

Occupational Lead Exposure

Occupational lead exposure limits should be lowered to 50 $\mu g/m^3$ in air and 40 $\mu g/dl$ in blood for males and females over the reproductive age and 30 **$\mu g/dl$** for women of child-bearing age.

Blood Lead Levels

The "intervention level" for blood-lead levels should be reduced to 15 $\mu g/dl$.

Monitoring Methods

Dustfall

An expanded monitoring network for lead in dustfall should be established or expanded at the provincial level.

Especially in areas around lead industries, provincial governments should conduct systematic data collection of dustfall and housedust lead levels and make the resulting data available to the general public within a reasonable time period.

Autobody Shops

Lead levels in soil and dust around automobile repair establishments should be monitored periodically.

Automobiles

An expanded system of monitoring pollution control equipment on vehicles should be established.

The Standard-Setting Process

Prevention

A policy of prevention of harm to the environment or human health should be adopted by regulatory agencies rather than a policy of reaction to problems after they arise.

Safety Margins

The setting of regulations should automatically incorporate a reasonably large safety margin to ensure prevention of harm to the most sensitive segments of the population.

Mandatory Review

Regulatory agencies should be required to conduct a mandatory review of all pollution standards and guidelines at five year intervals.

Unambiguous Controls

Control limits should be made directly applicable to individual point sources.

Decision-making

Public participation in decision- making on control limits and discharge permits should be sought and utilized.

The setting of standards should take into account the fact that some "classes" of pollutants have more severe effects (e.g., they are persistent or can cause irreversible damage) than others and, therefore, control strategies for them should be stricter.

Legal Considerations

Fines and Imprisonment

Violations of environmental regulations should involve larger fines and/or imprisonment of offenders.

The Burden of Proof

In cases of civil law, the burden of proof should fall on a producer or industry to prove that his products or activities are safe.

Legislative Changes

"Right-to-Know" legislation should be enacted in order for workers and the public to be assured of full access to information on substances and processes that may be hazardous to them.

Environmental protection and public health legislation should be amended to charge senior officials with a legal obligation to enforce these laws.

Enforceable decommissioning regulations, including the financial responsibility of the industry for decommissioning activities, should be developed in all provinces.

Other Considerations

Public Information

Provincial or municipal departments of health should investigate mechanisms for informing new residents in a lead "hotspot" area about the potential risks of lead exposure, particularly during renovation activities.

An environmental information office should be established to coordinate the availability of information from all levels of government.

Intervenor Funding

Where appropriate, funds should be made available for members of the public or groups acting in the public interest for presenting their position in legal or other situations.

Misfuelling

Government should work with the oil refining industry to bring about price equalization of leaded and unleaded fuel to eliminate the problem of misfuelling.

Appendix J

Regulations and Guidelines for Lead in Canada

The following tables list the federal and provincial regulations and guidelines for lead in Canada. Due to the wide variety of areas where lead is found, measured, and regulated, there may be additional guidelines and regulations in individual provinces or municipalities that are not included here. Information on specific guidelines or regulations can be obtained by contacting offices of individual federal or provincial ministries.

Footnotes for Table J.1

1. From operations involving the use of blast furnaces, cupolas, or reverberatory furnaces.
2. From operations involving the use of holding furnaces, pot furnaces, lead oxide production units or any other melting or refining operations or, scrap and material handling and storing, crushing, grinding, screening, conveying or casting, furnace tapping, furnace slagging, furnace cleaning, or any other source including building ventilation.
3. From the storage of lead-bearing scrap or material in or about a secondary lead smelter.
4. Paints, enamels or other liquid coating materials for use on furniture, toys and other articles intended for children and on pencils or artists' brushes.
5. Paints, enamels or other liquid coating materials for use on the interior or exterior surfaces of buildings, furniture or household products.
6. Total solid and dissolved lead in effluents from mines in existence before 1977 and all metal finishing plants.
7. Regulations apply to all new, re-opened (after 1977), or expanded (greater than 30%) mines other than gold mines.
8. British Columbia uses A, B, and C levels of pollution control objectives. A applies to all new or proposed discharges and is intended to provide adequate, long-term environmental protection. B applies to existing discharges which must be upgraded to B level and ultimately to A level by a staged program of improvement. C applies to all existing discharges which must be immediately upgraded to C level or within the shortest technically feasible time. C level is intended to provide adequate, short-term environmental protection. The time required for upgrading is determined on an individual basis and determined by the Director

of Pollution Control. None of these control objectives are directly enforceable but are used to establish levels for (enforceable) pollution control permits.
9. Standard cubic foot.
10. Alberta has not set a soil removal guideline but is currently (1986) developing guidelines. The Ontario guideline of 2600 ppm was used in 1983 as a site-specific level during the decommissioning of a lead-acid battery plant. Despite the adoption of this guideline, however, the more stringent level of 500 ppm was actually used during the clean-up.
11. Depending on site-specific variables, there are additional guidelines specifying the maximum accumulative addition of lead to the soil. Not more than one-third of the cumulative loading may be applied in a single application and sites previously treated with sludge may be re-treated after a period of three years has elapsed, if the available nitrogen levels in the soil have declined to less than 250 kg/ha in the top 150 cm depth of soil.
12. If objectives are exceeded, studies may be conducted by the Environmental Management Division and if waste effluents are the source, the Clean Environment Commission can apply legally binding emission orders on individual sources.
13. If irrigation water is sole source of water.
14. The guidelines also specify other variables related to metal availability (presence of nitrogen), maximum permissible lead content of soil before addition of sludge (60 ppm), and the maximum permissible lead content of the sludge (1100 ppm).

Sources: *Various governmental documents, and personal communication with various provincial governmental departments.*

Note: Detailed information on lead regulations and guidelines in the United States is available from the Environmental Protection Agency or State Environmental Agencies.

Table J.1

Regulations and Guidelines for Lead in the Environment

Federal

Controlling Authority of Jurisdiction	Lead Source	Lead Limit	Date Operative	Legislation and Current Status
Environment Canada	leaded gasoline	0.77 g/L	1976	*Clean Air Act* regulations
		0.29 g/L	1987	
		0.026 g/L (proposed)	1992	
	"lead-free" gasoline	0.013 g/L	1974	
Environment Canada and provincial environment ministries	stack emissions of particulates from secondary lead smelters	46 μg/m³ (63% lead)[1] 23 μg/m³ (63% lead)[2]	1976	
	particulate emissions[3] from secondary smelters	none allowed except as a result of handling		
Consumer and Corporate Affairs	paint	5000 ppm[4] 5000 ppm[5]	1970 1976	*Hazardous Products Act* regulations
	kettles	0.05 ppm (released into water)	1974	
	glazed ceramics	7 ppm (permissible acid soluble lead)	1971	
Health and Welfare	food	see Table J.2	1968	*Food and Drug Act* regulations

Health and Welfare and provincial environment ministries	drinking water	50 ug/L	early 1970's	guideline
Department of Fisheries and provincial environment ministries	lead in industrial effluents	1.5 mg/L[6] 0.2 mg/L (monthly mean) 0.3 mg/L (composite sample) 0.4 mg/L (grab sample)	1977[7]	guideline *Fisheries Act* regulations
Provincial				
British Columbia	ambient air around mining and smelting industries	1 to 2.5 μg/m³ (no averaging time specified)	1979	air quality objective
	ambient air around food processing and miscellaneous industries	A B C[8] 4 4 6 μg/m³/24 hours 2 2 3 μg/m³/year	1975	air quality objectives
	gaseous and particulate emissions	0.16 to 0.27 μg/mol 0.003 to 0.005 gm/SCF[9]	1979	control objectives
	for lead smelting and refining	0.5 kg/tonne lead produced or 1 lb/ton lead produced	1979	control objectives
	effluents to marine and fresh water	0.05 to 0.2 mg/L	1979	control objective

Alberta	stack emissions of particulates from secondary smelters	46 mg/m^3 (63% lead)[1] 23 mg/m^3 (63% lead)[2]	1978	*Clean Air Act* regulation
	particulate emissions[3] from secondary smelters	none allowed except as a result of handling		
	soil	2600 ppm	1983[10]	removal guideline
	sewer discharges	1.0 mg/L	1978	guideline
	sewage sludge	50 to 100 kg/ha[11]	1982	guideline
Saskatchewan	no lead-related industries and no guidelines or regulations for lead			
Manitoba	ambient air	5.0 $\mu g/m^3$/24 hours	1974	guideline
	drinking water	50 $\mu g/L$	1977	guideline[12]
	water for aquatic life	varies with water hardness	1977	guideline
	irrigation water	5.0 mg/L	1977	guideline[13]
	temporary irrigation	10.0 mg/L	1977	guideline
	livestock watering	5.0 mg/L	1977	guideline
	soil/sod	2600 ppm	1977	removal guideline
Ontario	ambient air	10 $\mu g/m^3$/half hour (Point of Impingement)	1974	*Environmental Protection Act* regulation
		5.0 $\mu g/m^3$/24 hours	1974	criterion
		3.0 $\mu g/m^3$/30 days (arithmetic mean)	1985	criterion

	dustfall	100 mg/m^2/30 days	1974	guideline
	soil	2600 ppm	1976	removal guideline
		600 ppm (considered "excessive")	1974	guideline
		500 ppm ("upper limit of normal" in urban surface soil)	1984	guideline
	unwashed plant foilage	150 ppm (considered "excessive")	1974	guideline
		100 ppm (considered "excessive")	1977	guideline
	sewer discharges	1.0-5.0 mg/L	1977	municipal sewer use by-laws
	sewage sludge application to agricultural land	90 kg/ha[14]	1986	guideline
Quebec	no information available			
New Brunswick	ambient air	5.0 $\mu g/m^3$/24 hours	1974	guideline
Nova Scotia	sewer discharges	1.0-5.0 mg/L	1977	municipal sewers use by-laws
Prince Edward Island	no information available			
Newfoundland	ambient air	10 $\mu g/m^3$/half hour	1981	*Department of Consumer Affairs and Environment Act* regulation
		5.0 $\mu g/m^3$/24 hours		criterion
		2.0 $\mu g/m^3$/30 days (geometric mean)		criterion
	water discharge	0.2 mg/L	1980	*Department of Consumer Affairs and Environment Act* regulation

Table J.2

Lead in Food Regulations

Food Type	Maximum Lead Level Allowed (ppm)	(ppb)
Citric acid	10	10,000
Tartaric acid	10	10,000
Cream of tartar	20	20,000
Sodium bicarbonate	5	5,000
Baking powder	10	10,000
Phosphoric acid	5	5,000
Calcium phosphate	5	5,000
Sodium, potassium and ammonium phosphates	5	5,000
Sodium and potassium nitrates	10	10,000
Sodium nitrite	20	20,000
Aluminum compounds	10	10,000
Marine and freshwater animal products	10	10,000
Liver	2	2,000
Fresh fruits	7	7,000
Fresh vegetables	2	2,000
Gelatin	7	7,000
Gelling agents except gelatin	20	20,000
Dried herbs and spices	10	10,000
Apple juice, cider, wine and beer	0.5	500
Fruit juices except apple juice	0.2	200
Beverages as consumed and bottled water other than natural mineral water	0.2	200
Tea	10	10,000
Edible bone meal	10	10,000
Fish protein	0.5	500

Source: ***Food and Drug Act,*** *R. S. C., 1970, c. F27, B. 15.001.*

Endnotes

Chapter 2

1. Environment Canada, "Socio-economic Impact Analysis of Lead Phase-down Control Options," Environmental Protection Service, February, 1984, p. 23.
2. National Health and Nutrition Examination Survey Series 11, No. 233, "Blood Lead Levels for Persons Ages 6 Months-74 Years: United States, 1976-80," United States Department of Health and Human Services/Public Health Service, DHHS Publication No. (PHS) 84-1683, 1984.

Chapter 3

1. United States Environmental Protection Agency (hereinafter referred to as USEPA), "Air Quality Criteria for Lead," EPA- 600/8-83-028B, Environmental Criteria and Assessment Office, Research Triangle Park, N.C., 1984, p. 1-126.
2. Ibid., p. 7-47.
3. J. Raloff, "Childhood Lead: Worrisome National Levels," *Science News,* 121 (6), February 6, 1982.
4. R. W. Elias, "Lead Exposures in the Human Environment," in K. R. Mahaffey (ed) *Dietary and Environmental Lead: Human Health Effects,* New York, Elsevier, 1985, pp. 79-107.
5. Ibid; and J. Pirkle, Centers for Disease Control, personal communication, March, 1986.
6. J. Roberts, K. R. Mahaffey, and J. L. Annest, "Blood Lead Levels in General Populations," in K. R. Mahaffey (ed) *Dietary and Environmental Lead: Human Health Effects,* New York, Elsevier, 1985, p. 361.
7. Use of a phosphate product should be infrequent and as limited as possible in order to prevent pollution through the sewage system of bodies of water, especially the Great Lakes.
8. Most lead particles (from industrial discharges, improper waste engine oil disposal, street dust runoff, and other miscellaneous sources) that enter the sewer system become concentrated in the sludge that settles at the bottom of the treatment tanks.
9. In Ontario, up to 90 kilograms of lead can be added to each hectare of soil. The maximum permissible level in soil is 60 ppm. Ontario Ministries of Agriculture and Food, Environment, and Health, "Guideline for Sewage Sludge Utilization on Agricultural Lands," January, 1986.
10. A. R. Bullen, "Getting the Lead Out," *Environmental Action,* March/April, 1986, p. 21.
11. Ibid. A 10% sodium sulphide solution can be obtained through surgical supply companies.

Chapter 4

1. H. A. Waldron and D. Stofen, *Sub-clinical Lead Poisoning,* London, Academic Press, 1974, p. 5.
2. Calculated on the basis of average annual consumption of 120,000 tonnes of lead in 1983 and 1984 (according to T. Yates, Cominco, Ltd, Toronto, personal communication, June, 1985) and a Canadian population of approximately 24 million. The per capita lead usage is probably similar in the U.S. and other industrialized nations.
3. M. Murozumi, T. J. Chow, and C. C. Patterson, "Chemical Concentrations of Pollutant Lead Aerosols, Terrestrial Dusts and Sea Salts in Greenland and Antarctic Snow Strata," *Geochim. Cosmochim. Acta,* 33, 1969, pp. 1247-1294.
4. G. B. Wiersma and C. Davidson, "Trace Metals in the Atmosphere of Remote Areas," in J. O. Nriagu and C. I. Davidson (eds) *Toxic Metals in the Air,* New York, Wiley, 1984.

Chapter 5

1. D. M. Settle and C. C. Patterson, "Lead in Albacore: Guide to Lead Pollution in Humans," *Science,* 207, 1980, pp. 1167-1176.
2. Environment Canada, "National Inventory of Sources and Releases of Lead (1982)," in Royal Society of Canada's Commission on Lead in the Environment, "Lead in Gasoline: A Review of the Canadian Policy Issue," 1985, p. 6.
3. C. Duncan, R. A. Kusiak, J. O'Heany, L. F. Smith, L. Spielberg, and J. Smith, "Blood Lead and Associated Risk Factors in Ontario Children, 1984," Summary and Conclusions of Technical Working Group Report, Ontario Ministries of Environment, Health, and Labour, 1985, p. 7.
4. Royal Society of Canada's Commission on Lead in the Environment, "Lead in Gasoline: A Review of the Canadian Policy Issue," 1985, p. 36.
5. Sewage sludge can be burned without releasing all of the lead contained in it if appropriate technology is used in the stack to trap lead particles. However, most sewage sludge incinerators do not have the technology in place to achieve this level of control.
6. R. W. Elias, "Lead Exposures in the Human Environment," in K. R. Mahaffey (ed) *Dietary and Environmental Lead: Human Health Effects,* New York, Elsevier, 1985, pp. 79-107; and J. Pirkle, Centers for Disease Control, personal communication, March, 1986.
7. C. C. Patterson, "An Alternative Perspective — Lead Pollution in the Human Environment: Origin, Extent, and Significance," in *Lead in the Human Environment,* National Academy of Sciences, Committee on Lead in the Human Environment, Washington, D.C., National Academy of Sciences, 1980, pp. 265-350; and J. Servant, "Atmospheric Trace Elements from Natural and Industrial Sources," University of London, Monitoring and Assessment Research Centre, London, 1982.
8. Environment Canada, "Ambient Air Particulate Lead Concentrations in Canada 1975-1983," Draft Report, 1985.

Chapter 6

1. D. Otto, V. Benignus, K. Muller, C. Barton, K. Seiple, J. Prah, and S. Schroeder, "Effects of Low to Moderate Lead Exposure on Slow Cortical Potentials in Young Children: Two Year Follow-up Study," *Neurobehav. Toxicol. Teratol.,* 4, 1982, pp. 733-737.
2. Centers for Disease Control, "Preventing Lead Poisoning in Young Children," United States Department of Health and Human Services/Public Health Service, Atlanta, Georgia, 1985, p. 1.
3. S. Piomelli, "The Effects of Low-level Lead Exposure on Heme Metabolism," in H. L. Needleman (ed) *Low Level Lead Exposure: The Clinical Implications of Current Research,* New York, Raven Press, 1980, p. 72.
4. Information in this section is drawn from USEPA, "Air Quality Criteria for Lead," EPA-600/8-83-028B, Environmental Criteria and Assessment Office, Research Triangle Park, N.C., 1984, pp. 12-13, to 12-49 and related sources.
5. 1,25-dihydroxyvitamin D.
6. J. Schwartz, H. Pitcher, R. Levin, B. Ostro, and A. L. Nichols, "Costs and Benefits of Reducing Lead in Gasoline: Final Regulatory Impact Analysis," EPA-230-05-85-006, Office of Policy Analysis, USEPA, Washington, D.C., February, 1985, pp. IV-5 to IV-6, IV-30 to IV-42; and USEPA, "Air Quality Criteria for Lead," EPA-600/8-83-028B, Environmental Criteria and Assessment Office, Research Triangle Park, N.C., 1984, pp. 1-139 to 1-144.
7. H. L. Needleman, C. Gunnoe, A. Leviton, R. Reed, H. Peresie, C. Maher, and P. Barrett, "Deficits in Psychologic and Classroom Performance of Children with Elevated Dentine Lead Levels," *New England Journal of Medicine,* 300 (13), 1979, pp. 689-695.
8. M. Rutter, "Raised Lead Levels and Impaired Cognitive/Behavioural Functioning: A Review of the Evidence," *Supp. Developmental Medicine and Child Neurology,* 22 (1), 1980, p. 21.
9. National Health and Nutrition Examination Survey Series 11, No. 233, "Blood Lead Levels for Persons Ages 6 Months-74 Years: United States, 1976-80," United States Department of Health and Human Services/Public Health Service, DHHS Publication No. (PHS) 84-1683, 1984.

Chapter 7

1. USEPA, "Air Quality Criteria for Lead," EPA-600/8-83- 028B, Environmental Criteria and Assessment Office, Research Triangle Park, N.C., 1984, p. 7-23.
2. Ibid., p. 7-68.
3. Ibid., p. 11-158.
4. B. Wallace and K. Cooper, *Lead, People, and the Environment,* Niagara Neighbourhood Association, Toronto, Ontario, 1985.
5. J. Kawasaki, Air Management Branch, Ontario Ministry of the Environment, memo to W.J. Hogg, Air Management Branch, March 12, 1973.
6. C. J. Macfarlane, Director, Air Management Branch, Ontario Ministry of the Environment, to W. Lachocki, May 3, 1973; W. J. Hogg, Air Management Branch, Ontario Ministry of the Environment, to South of King Residents' Association, July 27, 1973;

and W. J. Hogg, Air Management Branch, Ontario Ministry of the Environment, to Mrs. W. Lachocki, October 12, 1973.
7. W. Lachocki, Submission to special meeting of Toronto City Council, December 13, 1973, p.1.
8. Ontario Environmental Hearing Board, "Public Hearing on Lead Contamination in the Metropolitan Toronto Area," 1976.
9. Ibid, p. 20. A more stringent level of 1000 ppm had in fact been recommended in two large investigations conducted by the Ontario Ministry of the Environment (The Working Group on Lead and the Lead Data Analysis Task Force) and by the Institute for Environmental Studies at the University of Toronto.
10. C. C. Lax, "The Toronto Lead Smelter Controversy," in *Ecology versus Politics in Canada,* W. Leiss (ed), Toronto, University of Toronto Press, 1979, pp. 57-71.
11. Wallace and Cooper, op.cit., pp. 90-103.

Chapter 8

1. J. M. Ratcliffe, *Lead in Man and the Environment,* West Sussex England, Ellis Horwood Ltd, 1981, p. 149.
2. P. B. Hammond, "Metabolism of Lead," in J. J. Chisholm and D. M. O'Hara (eds) *Lead Absorption in Children,* Baltimore, Urban and Schwarzenberg, 1982, pp. 11-20.
3. Environment Canada, "Socio-economic Impact Analysis of Lead Phase-down Control Options," Environmental Protection Service, February, 1984, p. 21.
4. A. C. Chamberlain, "Prediction of Response of Blood Lead to Airborne and Dietary Lead from Volunteer Experiments with Lead Isotopes," *Proc. Royal Society of London,* B, 224, 1985, pp. 149-182.
5. USEPA, "Air Quality Criteria for Lead," EPA-600/8-83- 028B, Environmental Criteria and Assessment Office, Research Triangle Park, N.C., 1984, p. 1-68.
6. Ibid., p. 1-69 to 1-70.
7. Ibid., p. 1-71.
8. J. F. Jaworski, "Effects of Lead in the Environment: Quantitative Aspects," National Research Council of Canada, Report No. 16736, 1978, p. 456.
9. USEPA, op. cit., pp. 10-23.
10. K. R. Mahaffey, J. L Annest, J. Roberts, and R. S. Murphy, "National Estimates of Blood Lead Levels: United States, 1976-1980: Association with Selected Demographic and Socio-economic Factors," *New England Journal of Medicine,* 307, 1982, pp. 573-579.
11. USEPA, op. cit., p. 10-32.
12. Jaworski, op. cit., p. 377.
13. K. R. Mahaffey, "Relation between Quantities of Lead Ingested and Health Effects of Lead in Humans," *Pediatrics,* 59, 1977, pp. 448-456.
14. R. W. Dabeka, K. F. Karpinski, A. O. McKenzie, and C. D. Bajdik, "Survey of Lead, Cadmium, and Fluoride in Human Milk and Correlation of Levels with Environmental and Food Factors," *Food and Chemical Toxicology,* in press, 1986.

Chapter 9

1. J.I. Cheng, "A Blood Lead Survey of Southern Alberta," unpublished paper presented to Can. Public Health Assoc. Conference, Saskatoon, June 26, 1981.

2. National Health and Nutrition Examination Survey Series 11, No. 233, "Blood Lead Levels for Persons Ages 6 Months-74 Years: United States, 1976-80," United States Department of Health and Human Services/Public Health Service, DHHS Publication No. (PHS) 84-1683, 1984.
3. J. Pirkle, Centers for Disease Control, personal communication, March, 1986.
4. Although these percentages are the best available data, since they were obtained from children in Ontario, which is more urbanized and industrialized than much of the rest of Canada, they may be somewhat high for the country as a whole.
5. L. H. Hecker, H. E. Allen, B. D. Dinman, and J. V. Neel, "Heavy Metal Levels in Acculturated and Unacculturated Populations," *Arch. Environ. Health,* 29, 1974, p. 181- 185.
6. C. Poole, L. E. Smythe, and M. Alpers, "Blood Lead Levels in Papua New Guinea Children Living in a Remote Area," *The Science of the Total Environment,* 15, 1980, pp. 17-24.
7. P. C. Elwood, R. Blaney, R. C. Robb, A. J. Essex-Cater, B. E. Davies, and C. Toothill, "Lead Levels on Traffic-less Islands," *Journal of Epidemiology and Community Health,* 39, 1985, pp. 256-258.

Chapter 10

1. J. Schwartz, H. Pitcher, R. Levin, B. Ostro, and A. L. Nichols, "Costs and Benefits of Reducing Lead in Gasoline: Final Regulatory Impact Analysis," EPA-230-05-85-006, Office of Policy Analysis, USEPA, Washington, D. C., February, 1985, pp. VII-22 to VII-36.
2. Ibid, p. I-9.
3. Ibid, p. VII-3 to VII-14.
4. USEPA, "Air Quality Criteria for Lead," EPA-600/8-83- 028B, Environmental Criteria and Assessment Office, Research Triangle Park, N.C., 1984, pp. 5-10 to 5-14.
5. Motor Vehicle Manufacturers' Association, "Incentives for Proper Usage of Unleaded Fuel," January 11, 1984.
6. A. Laveskog, "Gasoline Additives: Past, Present, and Future," in P. Grandjean (ed) *Biological Effects of Organolead Compounds,* Boca Raton, Florida, CRC Press Inc., 1984, pp. 6-12.
7. S. Marsh, Suncor, Inc., personal communication, May, 1986.
8. J. Schwartz, Office of Policy Analysis, USEPA, "Change in Aromatics Emissions Due to Lead Phasedown," Memo to Docket EN-84-05, Office of Policy, Planning and Evaluation, USEPA, Washington, D.C., September 17, 1984; and Schwartz, et. al. op. cit., pp. II-2 to II-5 and VI-44 to VI-47. Misfuelling is expected to decrease with the lower lead standard because the greater refining required for leaded gasoline will likely bring about a reduction in the retail price differential between leaded and unleaded fuel thus reducing or eliminating the at-pump cost incentive to misfuel.
9. Environment Canada, "Socio-economic Impact Analysis of Lead Phase-down Control Options," Environmental Protection Service, February, 1984, p. 121.
10. Ibid.
11. Laveskog, op. cit.
12. D. R. Lynam, G. A. Hughmark, B. F. Fort, Jr. and C. A. Hall, "Blood Lead Concen-

trations and Gasoline Lead Usage;" and J. M. Pierrard, C. G. Pfeifer and R. D. Snee, "Assessment of Blood Lead Levels in the U.S.A. from NHANES *Data,* both papers presented at the *International Conference on Heavy Metals in the Environment,* Heidelberg, West Germany, Vol. I, September, 1983."
13. J. Rosenblatt, H. Smith, R. Royall, R. Little and J. R. Landis, "Report of the NHANES II Time Trend Analysis Review Group," USEPA, North Carolina, June, 15, 1983, pp. 12 and 14.
14. J. Schwartz, "The Relation Between Gas Lead and Blood Lead in Americans," unpublished paper presented at the International Conference on Heavy Metals in the Environment, Heidelberg, West Germany, September, 1983.
15. Environment Canada, op. cit. p. 22.
16. J. Ferguson, "Lead Industry Lobby earns Reputation for Toughness," *Globe and Mail,* Toronto, November 6, 1984.
17. This account of the 1920s controversy over leaded gasoline is drawn from the following article: D. Rosner and G. Markowitz, "A "Gift of God"?: The Public Health Controversy over Leaded Gasoline during the 1920s," *American Journal of Public Health,* 75 (4), 1985, pp. 344- 352.
18. Ibid, p. 350.
19. Ibid, p. 350.
20. Ibid, p. 348.
21. C. Duncan, R. A. Kusiak, J. O'Heany, L. F. Smith, L. Spielberg, and J. Smith, "Blood Lead and Associated Risk Factors in Ontario Children, 1984," Summary and Conclusions of Technical Working Group Report, Ontario Ministries of Environment, Health, and Labour, 1985.
22. As of January 1, 1987 when Canada lowers its standard to 0.29 g/L, the U.S. standard will be eleven times lower than the Canadian standard.
23. Schwartz, et. al. op. cit.

Chapter 11

1. T. Yates, Cominco, Ltd., personal communication, June, 1985.
2. Royal Society of Canada's Commission on Lead in the Environment, "Lead in Gasoline: A Review of the Canadian Policy Issue," 1985, p. 37.

Chapter 13

1. *Clean Air Act,* Leaded Gasoline Regulations, C.R.C., 1978.
2. *Canada Gazette,* Part II, Vol. 118, No.10, May 4, 1984, SOR/84-359.
3. J. Schwartz, H. Pitcher, R. Levin, B. Ostro, and A. L. Nichols, "Costs and Benefits of Reducing Lead in Gasoline: Final Regulatory Impact Analysis," EPA-230-05-85-006, Office of Policy Analysis, USEPA, Washington, D. C., February, 1985, p. E-2.
4. Environment Canada News Release, "Environment Minister to Ban Lead in Gasoline," PR-HQ-086-11, March, 1986. Despite the title of this release, several statements in the release, and additional material accompanying it, stating that lead would be "virtually disappearing" or "effectively eliminated" from gasoline by the end of 1992, this "ban" was described verbally by Mr. McMillan at the news conference as an intention to reduce the gasoline lead standard to the U.S. limit of 0.026 g/L. As of May, 1986, this limit has yet to be formally promulgated in the *Canada Gazette.*

5. C. Duncan, R. A. Kusiak, J. O'Heany, L. F. Smith, L. Spielberg, and J. Smith, "Blood Lead and Associated Risk Factors in Ontario Children, 1984," Summary and Conclusions of Technical Working Group Report, Ontario Ministries of Health, Environment, and Labour, 1985, p. 20. This report identifies urban Ontario children to have an average blood- lead level of about 12 ug/dl and the average for all children to be about 10 ug/dl. The average blood-lead level in United States urban children in 1983 is estimated to be about 10 ug/dl by J. Pirkle, Centers for Disease Control, personal communication, March, 1986.
6. Duncan, et. al., op. cit., p. 10.
7. Schwartz, et. al., op. cit., p. E-2.
8. Schwartz, et. al., op. cit.
9. Environment Canada, "Socio-economic Impact Analysis of Lead Phase-down Control Options," Environmental Protection Service, February, 1984, p. xi.
10. Royal Society of Canada's Commission on Lead in the Environment, "Lead in Gasoline: A Review of the Canadian Policy Issue," 1985, p. xiv.
11. Duncan, et. al., op. cit., p. 20.
12. As stated by federal Environment Minister, Tom McMillan in the Environment Canada News Release, in March, 1986, op. cit.
13. See discussion in Chapter 10 "The Leaded Gasoline Story" sub-section, "Alternatives to Lead."
14. Schwartz, et. al., op. cit., pp. I-15 to I-16 and II-3 to II-5.
15. Ibid, p. VIII-7 to VIII-26.
16. Environment Canada News Release, op. cit.
17. *Food and Drug Act,* R.S.C., 1970, c. F27, B. 15.001.
18. *Hazardous Products Act,* R.S.C., 1970, Vol. IV, c. H- 3, Schedule Part I.
19. *Hazardous Products Act,* Liquid Coating Materials Regulations, C.R.C., 1978.
20. Centers for Disease Control, "Preventing Lead Poisoning in Young Children," United States Department of Health and Human Services/Public Health Service, Atlanta, Georgia, 1985, p. 5.
21. For example: Ontario Regulation 536/81, Regulation respecting Lead — made under the *Occupational Health and Safety Act,* R.S.O., 1980, Chapter 321.
22. K. Bridbord, "Low-Level Lead Exposure in the Workplace," in H. L. Needleman (ed) *Low Level Lead Exposure: The Clinical Implications of Current Research,* New York, Raven Press, 1980, pp. 267-278.
23. W. Kuit, Cominco Ltd., presentation to the Royal Society of Canada's Commission on Lead in the Environment, April 26, 1985.
24. Ontario Regulation, 536/81, op. cit.
25. World Health Organization, *Recommended Health Based Limits in Occupational Exposure to Heavy Metals,* World Health Organization, Geneva, 1980, pp. 74-76.
26. International Lead Zinc Study Group, *Environmental and Health Controls on Lead,* London, England, Metro House 1985.
27. Ibid.
28. Ontario Ministy of Labour, "Proposed Regulation respecting Lead on Construction Projects made under the *Occupational Health and Safety Act,*" January 28, 1986.
29. S. Contenta, "Workers Dying as Safety Laws Ignored, Inspectors Charge," *Toronto Star,* May 21, 1986. This article points out that there are 207 inspectors responsible for about 150,000 workplaces under the Ministry of Labour's jurisdiction.

30. Ontario Ministry of the Environment, "List of Standards, Ambient Air Quality Criteria, Tentative Standards, Guidelines, and Provisional Guidelines for Air Contaminants," Emission Technology and Regulation Development Section, Air Resources Branch, January 2, 1985. A criterion of 3.0 ug/m^3 is the objective set for an arithmetic average of thirty daily air-lead measurements.
31. Ontario Regulation 872/74, Regulation 296 under the *Environmental Protection Act,* R.S.O., 1980.
32. K. H. Sharpe, G. W. O. Moss, G. J. Stopps, T. Anderson, L. Shenfeld, S. Linzon, G. S. Trivett, D. Bartkiw, D. J. Ogner, H. Nelson, and R. M. R. Higgin, "Studies of the Relationship of Environmental Lead Levels and Human Lead Intake," Ontario Ministry of the Environment, Working Group on Lead, 1974, p. 108.
33. B. Wallace and K. Cooper, *Lead, People, and the Environment,* Niagara Neighbourhood Association, Toronto, Ontario, 1985, pp. 152-157.
34. Sharpe et. al., op. cit., p. 108.
35. Ibid, p. 109.
36. Wallace and Cooper, op. cit., pp. 95-103.
37. Sharpe et. al., op. cit., p. 109.
38. Ontario Ministry of the Environment, Air Resources Branch, Phytotoxicology Section, Report ARB-075-85 Phyto, 1985. This soil-lead increase was from a level of 650 ppm in 1983 to 970 ppm in 1984.
39. G. J. Stopps, "Bioavailability of Lead in Soil near a Secondary Lead Smelter," unpublished article, University of Toronto, 1980.
40. Centers for Disease Control, op. cit., p. 7.
41. I. Thornton, E. Culbard, S. Moorcroft, J. Watt, M. Wheatley, and M. Thompson, "Metals in Urban Dusts and Soils," *Env. Tech. Letters,* 6, 1985, pp. 137-144. The authors state that the internal "enrichment" may reflect internal sources, such as paint, selective movement indoors of fine (metal-rich) particles and/or the selective removal of particles low in metals during house cleaning.
42. This collector (Ontario Ministry of the Environment Station No. 31054) is located several blocks away from an inadequately-controlled lead smelter. Collectors located close to this smelter are obviously affected by its fugitive emissions since the government guideline for lead in dustfall is regularly exceeded. However, these emissions appear to extend only a very short distance out from the smelter and the collector located farther away likely reflects a fairly typical inner-city lead fallout rate.
43. J. Raloff, "Childhood Lead: Worrisome National Levels," *Science News,* 121 (6), February 6, 1982.
44. Environment Canada, op. cit., p. 21.

Chapter 14

1. Problems with the location of air-lead monitors include such things as access to a power source, access to private property, and security from vandalism.
2. C. Duncan, Ontario Ministry of the Environment, personal communication, May, 1986.
3. *Clean Air Act,* Secondary Lead Smelter National Emission Standards Regulations, C.R.C. 1978.

4. Ontario Ministry of the Environment, "Air Pollution General Regulation Workshop," November, 1985; and "Proceedings of Air Pollution General Regulation Workshop, Nov. 14/15th, 1985," March, 1986.
5. Ibid.
6. R.B. Gibson, "Control Orders and Industrial Pollution Abatement in Ontario," Canadian Environmental Law Research Foundation, 1983, p. 7.

Chapter 15

1. Ontario Ministry of the Environment, "Guidelines for Decommissioning (Shutdown) of Major Industrial Sites in Ontario," 1984.
2. Monenco Consultants Ltd., "Guide to the Environmental Aspects of Decommissioning of Industrial Sites," Calgary, Alberta, Monenco Consultants, 1985, pp. 4-8 to 4-9.
3. Ibid, pp. 3-91 to 3-97.
4. S. Winterton, Pollution Probe, Toronto, personal communication, April 17, 1986.
5. Environment Canada, press release, June, 1985.
6. J. O. Nriagu, Environment Canada, personal communication, April 15, 1986.
7. Centers for Disease Control, "Preventing Lead Poisoning in Young Children," United States Department of Health and Human Services/Public Health Service, Atlanta, Georgia, 1985, p. 7.
8. A. S. MacPherson, "South Riverdale Lead Control Issues," Report to the Local Board of Health, Toronto, February 20, 1985.
9. J. W. Assink and H. J. van Veen, "Extractive Cleaning of Metal Contaminated Soil and Groundwater," *Hoofdgroep Maatschappelijke Technologie,* 1985, Table 14-1.

Bibliography

Assink, J. W., and van Veen, H. J., 1985. "Extractive Cleaning of Metal Contaminated Soil and Groundwater," *Hoofdgroep Maatschappelijke Technologie.*

Billick, H. H., 1980. "Lead: A Case Study in Interagency Policy-making," *Environmental Health Perspectives,* 42:73- 79.

Bridbord, K., 1980. "Low-level Lead Exposure in the Workplace." In *Low Level Lead Exposure: The Clinical Implications of Current Research,* ed. H. L. Needleman. New York, Raven Press.

Bullen, A. R., 1986. "Getting the Lead Out." *Environmental Action,* March/April.

Canada. Department of National Health and Welfare, 1982. "Human Exposure to Environmental Lead." Report prepared for the Department of the Environment, Monitoring and Criteria Division, Bureau of Chemical Hazards, Ottawa.

Canadian Broadcasting Corporation, 1974. CBC Radio Public Affairs. Transcript of "As It Happens" segment "Dying of Lead," broadcast January 29, 1974.

Carnow, B. W., P. S. Levy, and R. A. Wadden, 1974. "Report to the Toronto Board of Health: Assessment of Effect of Environmental Lead."

Chamberlain, A. C., 1985. "Prediction of Response of Blood Lead to Airborne and Dietary Lead from Volunteer Experiments with Lead Isotopes," *Proc. Royal Society of London,* B, 224.

Cheng. J. I., 1981. "A Blood Lead Survey of Southern Alberta." Unpublished paper presented to Can. Public Health Assoc. Conference, Saskatoon, June 26, 1981.

Contenta, S., 1986. "Workers Dying as Safety Laws Ignored, Inspectors Charge," *Toronto Star,* May 21, 1986.

Dabeka, R. W., K. F. Karpinski, A. O. McKenzie, and C. D. Bajdik, 1986. "Survey of Lead, Cadmium, and Fluoride in Human Milk and Correlation of Levels with Environmental and Food Factors," *Food and Chemical Toxicology,* in press.

Duncan, C., R. A. Kusiak, J. O'Heany, L. F. Smith, L. Spielberg, and J. Smith, 1985. "Blood Lead and Associated Risk Factors in Ontario Children, 1984," Technical Working Group Report, Ontario Ministries of Environment, Health, and Labour.

Elias, R. W., 1985. "Lead Exposures in the Human Environment." In *Dietary and Environmental Lead: Human Health Effects,* ed. K. R. Mahaffey. New York, Elsevier.

Elwood, P. C., R. Blaney, R. C. Robb, A. J. Essex-Cater, B. E. Davies, and C. Toothill, 1985. "Lead Levels on Traffic- less Islands," *Journal of Epidemiology and Community Health,* 39:256-258.

Environment Canada, 1982. *Lead-free and Leaded Gasoline Enforcement Programs, 1974-80.* Air Pollution Directorate, Economic and Technical Review Report EPS 3-AP-82-3. Ottawa.

Environment Canada, 1983. "National Inventory of Sources and Releases of Lead (1978)." Environmental Protection Service, Economic and Technical Review Report EPS 3-EP-83-6.

Environment Canada, 1984. "Socio-economic Impact Analysis of Lead Phase-down Control Options." Environmental Protection Service. Ottawa.

Environment Canada, 1985. "National Inventory of Sources and Releases of Lead (1982)." In Royal Society of Canada's Commission on Lead in the Environment, "Lead in Gasoline: A Review of the Canadian Policy Issue."

Environment Canada, 1985. "Ambient Air Particulate Lead Concentrations in Canada 1975-1983," draft report.

Environment Canada, 1986. "Environment Minister to Ban Lead in Gasoline." News release PR-HQ-086-11, March.

Facchetti, S., 1979. "Isotope Study of Lead in Petrol." In *International Conference: Management and Control of Heavy Metals in the Environment,* September 1979, pp. 95-102. London and Edinburgh, CEP Consultants, Ltd.

Facchetti, S., and F. Geiss, 1982. *Isotopic Lead Experiment: Status Report,* Luxembourg: Commission of the European Communities, Publication No. EUR 8352, EN.

Ferguson, J., 1984. "Lead Industry Lobby earns Reputation for Toughness," *Globe and Mail,* Toronto, November 6, 1984.

Fischbein, A., 1983. "Environmental and Occupational Lead Exposure." In *Environmental and Occupational Medicine,* ed. W. N. Rom, pp. 433-447. Boston, Little, Brown & Co.

Gibson, R. B., 1983. "Control Orders and Industrial Pollution Abatement in Ontario." Toronto, Canadian Environmental Law Research Foundation.

Great Britain. Department of Health and Social Security, 1980. "Lead and Health." Report of a DHSS Working Party on Lead in the Environment. London, H. M. S. O.

Hammond, P. B., 1982. "Metabolism of Lead." In *Lead Absorption in Children,* eds. J. J. Chisholm and D. M. O'Hara, pp. 11-20. Baltimore, Urban and Schwarzenberg.

Hecker, L. H., H. E. Allen, B. D. Dinman, and J. V. Neel, 1974. "Heavy Metal Levels in Acculturated and Unacculturated Populations," *Arch. Environ. Health,* 29:181-185.

International Lead-Zinc Study Group, 1985. *Environmental and Health Controls on Lead.* London, Metro House.

Jaworski. J. F., 1978. "Effects of Lead in The Environment; Quantitative Aspects." National Research Council of Canada, Report No. 16736.

Laveskog, A., 1984. "Gasoline Additives: Past, Present, and Future." In *Biological Effects of Organolead Compounts,* ed P. Grandjean, pp. 6-12. Boca Raton, Florida, CRC Press Inc.

Lax. C. C., 1979. "The Toronto Lead Smelter Controversy." In *Ecology versus Politics in Canada,* ed W. Leiss, pp. 57- 71. Toronto, University of Toronto Press.

Lachocki, W., 1973. Submission to Toronto City Council, December 13, 1973.

Lynam, D. R., G. A. Hughmark, B. F. Fort, Jr., and C. A. Hall, 1983. "Blood Lead Concentrations and Gasoline Lead Usage." In *International Conference on Heavy Metals in the Environment,* Vol. I. Heidelberg, West Germany.

MacPherson, A. S., 1985. "South Riverdale Lead Control Issues." Report to the Toronto Board of Health. Toronto, February 20, 1985.

Mahaffey, K. R., 1977. "Relation between Quantities of Lead Ingested and Health Effects of Lead in Humans," *Pediatrics,* 59:448-456.

Mahaffey, K. R., J. L. Annest, J. Roberts, and R. S. Murphy, 1982. "National Estimates of Blood Lead Levels: United States, 1976-1980: Association with Selected Demographic and Socio-economic Factors," *New England Journal of Medicine,* 307:573-579.

Martin, C. B., and P. C. Kupa, 1977. "The Rationale, Methodology and Administration used in Ontario to Determine Ambient Air Objectives and Emission Standards." Paper presented at the 70th Annual Meeting of the Air Pollution Control Association, Toronto, June 19-24, 1977.

Monenco Consultants Ltd., 1985. "Guide to the Environmental Aspects of Decommissioning of Industrial Sites." Calgary, Alberta, Monenco Consultants.

Motor Vehicle Manufacturers' Association, 1984. "Incentives for Proper Usage of Unleaded Fuel." Paper dated January 11, 1984.

Murozumi, M., T. J. Chow, and C. C. Patterson, 1969. "Chemical Concentrations of Pollutant Lead Aerosols, Terrestrial Dusts, and Sea Salts in Greenland and Antarctic Snow Strata," *Geochim. Cosmochim. Acta,* 33:1247-1294.

National Academy of Sciences. Commission on Natural Resources, 1980. *Lead in the Human Environment.* Washington, D. C., National Academy of Sciences.

National Health and Nutrition Examination Survey Series 11, No. 233, 1984. "Blood Lead Levels for Persons Ages 6 Months- 74 Years: United States, 1976-80." United States Department of Health and Human Services/Public Health Service, DHHS Publication No. (PHS) 84-1683.

Needleman, H. L., C. Gunnoe, A. Leviton, R. Reed, H. Peresie, C. Maher, and P. Barrett, 1979. "Deficits in Psychologic and Classroom Performance of children with Elevated Dentine Lead Levels," *New England Journal of Medicine,* 300:689-695.

Nriagu, J. O., 1983. "Saturnine Gout among Roman Aristocrats: Did Lead Poisoning contribute to the Fall of the Empire?" *New England Journal of Medicine,* 308:660-663.

Nutrition Foundation, Inc., 1982. "Assessment of the Safety of Lead and Lead Salts in Food: A Report of the Nutrition Foundation Expert Advisory Committee." Washington, D. C., The Nutrition Foundation.

Ontario Environmental Hearing Board, 1976. "Public Hearing on Lead Contamination in the Metropolitan Toronto Area."

Ontario Ministries of Agriculture and Food, Environment, and Health, 1986. "Guideline for Sewage Sludge Utilization on Agricultural Lands."

Ontario Ministry of the Environment, 1984. "Guidelines for Decommissioning (Shutdown) of Major Industrial Sites in Ontario."

Ontario Ministry of the Environment, 1985. Report ARB-075-85 Phyto. Air Resources Branch, Phytotoxicology Section.

Ontario Ministry of the Environment, 1985. "List of Standards, Ambient Air Quality Criteria, Tentative Standards, Guidelines, and Provisional Guidelines for Air Contaminants." Emission Technology and Regulation Development Section, Air Resources Branch, January 2, 1985.

Ontario Ministry of the Environment, 1985. "Air Pollution General Regulation Workshop," November, 1985.

Ontario Ministry of the Environment, 1986. "Proceedings of Air Pollution General Regulation Workshop, November 14/15th, 1985," March, 1986.

Ontario Ministry of Labour, 1986. "Proposed Regulation respecting Lead on Construction Projects made under the *Occupational Health and Safety Act,*" January 28, 1986.

Otto, D., V. Benignus, K. Muller, C. Barton, K. Seiple, J. Prah, and S. Schroeder, 1982. "Effects of Low to Moderate Lead Exposure on Slow Cortical Potentials in Young Children: Two Year Follow-up Study," *Neurobehav. Toxicol. Teratol.,* 4:733-737.

Patterson, C. C., 1980. "An Alternative Perspective — Lead Pollution in the Human Environment: Origin, Extent, and Significance." In *Lead in the Human Environment,* National Academy of Sciences, Committee on Lead in the Human Environment, pp. 265-350. Washington, D. C., National Academy of Sciences.

Pierrard, J. M., C. G. Pfeifer, and R. D. Snee, 1983. "Assessment of Blood Lead Levels in the U.S.A. from NHANES II Data." In *International Conference on Heavy Metals in the Environment,* Vol. I. Heidelberg, West Germany.

Piomelli, S., 1980. "The Effects of Low-level Lead Exposure on Heme Metabolism." In *Low Level Lead Exposure: The Clinical Implications of Current Research,* ed. H. L. Needleman, p. 72. New York, Raven Press.

Poole, C., L. E. Smythe, and M. Alpers, 1980. "Blood Lead Levels in Papua New Guinea Children Living in a Remote Area," *The Science of the Total Environment,* 15:17-24.

Raloff, J., 1982. "Childhood Lead: Worrisome National Levels," *Science News,* 121 (6), February 6, 1982.

Ratcliffe, J. M., 1981. *Lead in Man and the Environment.* West Sussex, England, Ellis Horwood Ltd.

Roberts, J., K. R. Mahaffey, and J. L. Annest, 1985. "Blood Lead Levels in General Populations." In *Dietary and Environmental Lead: Human Health Effects,* ed K. R. Mahaffey. New York, Elsevier.

Roberts, T. M., and T. C. Hutchinson, 1974. "Report on Lead in Soil from the Railway Embankment between Strachan and Bathurst Streets," Institute for Environmental Studies, University of Toronto, March 27, 1974.

Robertson, H. R., D. A. Chant, and F. A. DeMarco, 1974. "Effects on Human Health of Lead from the Environment." Toronto, Queen's Park.

Robertson, T. M., J. J. Pacija, T. C. Hutchinson, R. E. Jervis, A. Chattopadhyay, J. C. Van Loon, and F. Huhn, 1974. "Lead Contamination around two Secondary Lead Smelters in Downtown Toronto — Estimation of ongoing Pollution and Accumulation by Humans," Institute for Environmental Studies, University of Toronto, October, 1974.

Rosenblatt, J., H. Smith, R. Royall, R. Little, and J. R. Landis, 1983. "Report of the NHANES II Time Trend Analysis Review Group," USEPA, North Carolina, June 15, 1983.

Rosner, D., and G. Markowitz, 1985. "A "Gift of God?": The Public Health Controversy over Leaded Gasoline during the 1920s," *American Journal of Public Health,* 75 (4):344- 352.

Royal Society of Canada's Commission on Lead in the Environment, 1985. "Lead in Gasoline: A Review of the Canadian Policy Issue."

Rutter, M., 1980. "Raised Lead Levels and Impaired Cognitive/Behavioural Functioning: A Review of the Evidence," *Supp. Developmental Medicine and Child Neurology,* 22 (1):1-26.

Savan, B., 1986. "Sleazy Science," *Alternatives,* 13 (2):11-17.

Schwartz, J., 1983. "The Relation between Gas Lead and Blood Lead in Americans." Unpublished paper presented at the International Conference on Heavy Metals in the Environment. Heidelberg, West Germany, September, 1983.

Schwartz, J., 1984. "Change in Aromatics Emissions due to Lead Phasedown." Memo to Docket EN-84-05, Office of Policy, Planning and Evaluation, USEPA, Washington, D. C., September 17, 1984.

Schwartz, J., H. Pitcher, R. Levin, B. Ostro, and A. L. Nichols, 1985. "Costs and Benefits of Reducing Lead in Gasoline: Final Regulatory Impact Analysis." EPA-230-05-85- 006, Office of Policy Analysis, USEPA, Washington, D. C., February, 1985.

Science Council of Canada, 1977. "Policies and Poisons: The Containment of Long-term Hazards to Human Health in the Environment and the Workplace." Science Council of Canada Report No. 28, pp. 115-129.

Servant, J., 1982. "Atmospheric Trace Elements from Natural and Industrial Sources." London, University of London, Monitoring and Assessment Research Centre.

Settle, D. M., and C. C. Patterson, 1980. "Lead in Albacore: Guide to Lead Pollution in Humans," *Science,* 207:1167- 1176.

Sharpe, K. H., G. W. O. Moss, G. J. Stopps, T. Anderson, L. Shenfeld, S. Linzon, G. S. Trivett, D. Bartkiw, D. J. Ogner, H. Nelson, and R. M. R. Higgin, 1974. "Studies of the Relationship of Environmental Lead Levels and Human Lead Intake." Toronto, Ontario Ministry of the Environment, Working Group on Lead.

Stopps, G. J., 1980. "Bioavailability of Lead in Soil near a Secondary Lead Smelter." Unpublished report, Toronto, University of Toronto.

Thornton, I., E. Culbard, S. Moorcroft, J. Watt, M. Wheatley, and M. Thompson, 1985. "Metals in Urban Dusts and Soils," *Env. Tech. Letters,* 6:137-144.

USEPA, 1984. "Air Quality Criteria for Lead." EPA-600/8-83- 028B, Environmental Criteria and Assessment Office, Research Triangle Park, N. C.

Waldron, H. A., and D. Stofen, 1974. *Sub-clinical Lead Poisoning,* London, Academic Press.

Wallace, B., and K. Cooper, 1985. *Lead, People, and the Environment.* Toronto: Niagara Neighbourhood Association.

Wiersma, G. B., and C. I. Davidson, 1984. "Trace Metals in the Atmosphere of Remote Areas." In *Toxic Metals in the Air,* eds J. O. Nriagu and C. I. Davidson. New York, Wiley.

World Health Organization, 1977. *Environmental Health Criteria 3 — Lead.* Task Group on Environmental Health Criteria for Lead, 1977. Geneva, World Health Organization.

World Health Organization, 1980. *Recommended Health-based Limits in Occupational Exposure to Heavy Metals,* pp. 74-76. Geneva, World Health Organization.

Glossary

ALA	aminolevulinic acid, an amino acid; as ALA-D decreases, ALA increases.
ALA-D	delta-aminolevulinic acid dehydrase, an enzyme that breaks down one kind of amino acid.
allowable level	refers to the amount of lead in blood that does not result in any medical action; considered by some as a "safe" level; for children, the medical profession generally uses 20 to 25 μg/dl, although others have recommended levels as low as 10 to 15 μg/dl.
alloy	a metal made by mixing and fusing two or more metals, or a metal and a non-metal; for example, brass is an alloy of copper and zinc.
ambient air	any unconfined portion of the atmosphere; the outside air.
amino acids	the building blocks for protein construction.
anemia	an insufficient amount of hemoglobin or of red blood cells in the blood.
arithmetic average	the common average obtained by dividing the sum of a set of values by the number of values in the set.
aromatics	a group of hydrocarbons found in crude oil characterized by distinctive, often pleasant, aromas and based on a unit of six carbons.
baghouse	an air pollution control device for removing particles from a gas stream by filtration with fabric bags.
biochemical risk level	refers to the amount of lead in blood that is capable of causing changes inside body cells and systems but that is not enough to cause overt health problems (i.e., recognizable "sickness"); usually set at 10 to 15 μg/dl for children.
catalytic converter	an air pollution control device for motor vehicles that is ruined by lead-containing gasoline (see text for two main types).
CDC	the United States Centers for Disease Control in Atlanta, Georgia.
central nervous system	the brain and spinal cord; see "nervous system."
chelation	a medical procedure used to remove lead from bone and soft tissues.
cognitive	having to do with factual knowledge and understanding; cognition is the mental process by which knowledge is acquired.
concentration	the amount of a substance in a specified volume of air or liquid.
criterion, criteria	an unforceable, maximum amount of a substance (substances) in some medium, such as air or soil.
cupellation	the process of separating silver from base metals, including lead.
decommission	the process of shutting down an industrial facility.

defendant	the person against whom a legal action is brought; the one charged with an illegal act.
detoxification	the removal or neutralization of a toxic substance or poison from or in a person or substance.
dispersion	the act or process of scattering in various directions or spreading out over a wide area.
EEG	electroencephalogram, a measurement of electrical activity in the brain.
electro-physiology	the study of electrical events that occur in the normal functioning of a living thing.
elevated level	refers to the amount of lead in blood that can cause health problems; an excessive absorption of lead; set in 1985 at 25 μg/dl for children.
emissions	substances that are given off, discharged, or that escape from a mobile or stationary source.
endocrine system	the system of hormone secretion from glands in various parts of the body.
EP	erythrocyte protoporphyrin, a substance in red blood cells that increases when the amount of lead in the blood increases.
erythrocytes	red blood cells.
exceedance	a measurement term used in monitoring environemtal pollutants (such as lead in air) that refers to an amount or concentration larger than the accepted standard, criterion, or guideline.
externalities	a term used by economists to describe those costs of a product or process not paid by individual producers or consumers but by society as a whole.
fatal	causing, or being capable of causing death.
fetus, fetal	in humans, the unborn child in the uterus from the third month to birth.
fugitive emissions	emissions of pollutants that arise from sources (e.g., yards, work piles, windows, vents, doorways, etc.) other than smoke stacks or designed emission points.
gastrointestinal tract	the stomach and intestines through which food passes.
geometric average	average for a set of numbers computed by taking the nth root of the product of n numbers; it is generally smaller than the arithmetic average
GI tract	gastrointestinal tract.
globin	the protein part of hemoglobin.
guideline	an unforceable, recommended, maximum level of a substance.
hazardous	a characteristic of a substance indicating that it can harm, contaminate, or kill living organisms because it is ignitable, corrosive, reactive, or toxic.
heavy metal	an element of the metal class whose density is at least five times greater than water.

heme the iron-containing, non-protein part of hemoglobin.

hemoglobin the iron and protein-containing pigment in red blood cells that carries oxygen from the lungs and carbon dioxide from the tissues.

hypertension a condition in which a person has a higher blood pressure than is considered normal.

immunoregulatory cells cells involved in the regulation of actions taken by the body to protect a person from disease.

inertia a tendency to remain in the same state and not change.

ingest to take into the body by swallowing.

inhale to take into the body by breathing.

inorganic refers to substances that are derived from minerals and do not contain carbon; does not include animal or vegetable substances.

insoluble the characteristic of being unable to be dissolved in water or another liquid.

intervention level refers to the amount of lead in blood that is used by some health departments as an "alert" signal indicating that the amount of absorbed lead is approaching the elevated level; usually set at 20 μg/dl for children.

I.Q. Intelligence Quotient, a number used to describe a person's relative intelligence in terms of certain thinking skills.

isomerization alteration of the internal structure of chemical substances. For example, isobutane is an isomer of butane that gives a high octane level to gasoline.

isopleth a line on a map or chart connecting places sharing some feature in common, such as the amount of lead in the soil.

"knocking" the premature combustion of the air-fuel mixture in an engine cylinder.

mean the arithmetic average value of a set of numbers.

median a type of average determined by selecting the middle number in a set of numbers arranged in order of size.

metabolism the process of building up food into living matter and of breaking down living matter into simpler substances.

misfuelling the use of leaded gasoline in a vehicle intended to use unleaded gasoline.

MMT methyl cyclopentadienyl manganese tricarbonyl, a possible additive to gasoline that performs essentially the same functions as does lead in gasoline.

monitoring the measurement of specific properties at some point to detect changes that might be due to some particular operation or facility.

neurotransmitter a chemical substance that transmits information from one nerve cell to another.

NA not available.

nervous system a system of extremely delicate nerve cells elaborately interlaced with each other, including the brain, spinal cord, sense organs (e.g., eye, ear), and peripheral nerves.

NHANES-II	the second United States National Health and Nutrition Examination Survey conducted from 1976 to 1980.
nucleic acids	an important group of substances essential to life, found especially in the nuclei of cells; includes DNA that contains the genetic codes determining heredity and RNA that aids in the formation of proteins and that is involved in the transfer of "messages" at the cellular level.
organic	refers to living organisms; in chemistry, refers to any compound containing carbon.
particulates	finely divided solid or liquid particles in the air or in an industrial emission.
per capita	for each person.
peripheral nervous system	the system composed of special sense organs, control of involuntary body functions(e.g., heart rate, glands), and nerves in the torso and limbs; see "nervous system."
persistency	having a long-lasting or permanent existence or quality.
pH	a symbol used to express relative acidity or alkalinity; the pH scale ranges from 1 (very acidic) to 14 (very aklaline); a pH of 7 (e.g., distilled water) at the centre of the scale is neutral (non-acid and non-alkaline).
Py-5-N	pyrimidine-5'-nucleotidase, an enzyme present in blood serum that breaks down nucleic acids.
phytotoxicology	the study of toxic effects on vegetation.
pica	the habit of eating non-food substances including peeling paint, plaster, putty, and dirt or soil.
placenta	the organ by which the fetus is attached to the wall of the womb and nourished.
plaintiff	the person or group who initiates a lawsuit against another party for a violation of law.
point of impingement	that environmental location touched by an industrial discharge and capable of being damaged by the discharge.
point source	a specific location from which pollutant emissions enter the environment; e.g., a lead smelter.
quantifiable	to be able to express something in terms of a numerical value.
re-entrain	the re-entry of something into the current environmental flow; e.g., lead in soil can be re-entrained into air lead particulates by the wind.
sapa	a sweet, but poisonous alternative to sugar produced during the time of the Roman Empire by boiling grape juice in a lead or lead-lined pot.
screening a population	obtaining a physical measurement (e.g., blood-lead level) on a sample of people from a population group to identify those who have high levels on that measurement.
secondary lead smelter	a plant or factory in which lead-bearing scrap or materials, other than lead-bearing concentrates derived from a mining operation, are processed by metallurgical or chemical processes into refined lead, lead alloys, or lead oxide.
sediment	any matter that settles to the bottom of a liquid; e.g., the mud at the bottom of a river or lake.

serum	the clear, pale-yellow, watery part of blood.
smelter	a place where ore is melted to get the metal out of it.
soft tissues	those parts of the body that are neither mineralized (e.g., teeth or bone) nor liquid (e.g., blood); includes organs, skin, muscles, and glands.
stack emissions	emissions of pollutants arising from a specific discharge stack.
standard	the maximum allowable level of a substance in some environmental medium, such as air or water, or in a consumer product or discharge stack, as specified in a legally enforceable regulation.
suburban	the residential sections, districts, towns, or villages on the outskirts of a city.
susceptibility	having little or low resistance to a disease or the harmful effects of a pollutant.
TEL	the abbreviation for tetraethyl lead, a common form of gasoline lead additive
threshold	the minimum level of a substance that is capable of producing an effect.
ton	a "short" ton; equivalent to 2000 pounds.
tonne	one metric tonne; equivalent to 1000 kilograms or 2204.6 pounds.
toxic	poisonous, potentially harmful to the health and/or reproductive ability of organisms.
urban	having to do with cities or large towns.
valve seat recession	abrasive and adhesive wear that can erode a car engine's valve-seats, requiring major engine repairs.
zinc protoporphyrin	a naturally occurring derivative of hemoglobin.

Index